THE SCIENCE OF

COMMUNICATING
SCIENCE

THE ULTIMATE GUIDE

DR CRAIG CORMICK

CSIRO

PUBLISHING

CABI

A catalogue record for this book is available from the National Library of Australia and from the British Library, London, UK.

ISBN: 9781486309818 (pbk)
ISBN: 9781486309825 (epdf)
ISBN: 9781486309832 (epub)

Published exclusively in all formats throughout the world (excluding in Europe, Middle East, Africa and Asia) by:

CSIRO Publishing
Locked Bag 10
Clayton South VIC 3169
Australia

Telephone: +61 3 9545 8400
Email: publishing.sales@csiro.au
Website: www.publish.csiro.au

Published exclusively in print only (Europe, Middle East, Africa and Asia) by CABI.

ISBN 9781789245141

CABI	CABI
Nosworthy Way	745 Atlantic Avenue
Wallingford	8th Floor
Oxfordshire OX10 8DE	Boston, MA 02111
UK	USA
Tel: +44 (0)1491 832111	Tel: +1 (617)682-9015
Fax: +44 (0)1491 833508	E-mail: cabi-nao@cabi.org
Email: info@cabi.org	
Website: www.cabi.org	

Set in 11/13 Adobe Garamond Pro and Myriad Pro
Edited by Karen Pearce
Cover design by Astred Hicks, Design Cherry
Typeset by Desktop Concepts Pty Ltd, Melbourne
Printed in the USA by Integrated Books International

Mar20_02

Contents

Acknowledgements

A very special thanks to all those who cast their eagle-eyes over the draft manuscript and provided insightful comments including Andrew Maynard, Brian Zikmund-Fisher, Elyse Aurbach, Susannah Eliott and Anna-Maria Arabia. A very special thank you to Katie Prater who went above and beyond in helping make this the book it now is.

Introduction

In reading the many hundred journal articles and media reports and surveys and blogs that informed the content of this book, one quote stood out that I kept aside for the introduction, as I think it best captured what I was trying to achieve.

Professor Brigitte Nerlich of Nottingham University in the UK wrote in a blog post entitled 'Science communication: What was it, what is it, and what should it be?':

> … a whole academic industry has begun to flourish that is supposed to tell scientists what to communicate, how to communicate and for what reasons to communicate. Research into these matters has proliferated (and I have contributed to this proliferation). Unfortunately, the results of that research are largely published in places and in languages that scientists don't visit and don't really understand. As a result, there is some estrangement between those who still communicate and those who want to tell them how to do it.

In short, that's a problem that we need to do something about. What is the point of having all this great research into how to better communicate science if it is largely inaccessible?

Increasingly scientists recognise the call made by former head of the US National Oceanographic and Atmospheric Administration (NOAA), Jane Lubchenco, to acknowledge there is a social contract to communicate science with society – but how to do that?

We have all these busy people doing the best they can in their fields, but who are far too time poor to hunt around all over the place and find those pearls of wisdom and research summaries that might help them better communicate science.

Somebody ought to do something about it, yeah?

Dr Craig Cormick

THE GROUND RULES

1

What makes good science communication? (in fewer than 280 characters)

'Somewhere something incredible is waiting to be known.'

– Carl Sagan, astronomer and science communicator

When I am asked what makes good science communication, I say, **'Just three things really – know your audience, and tell a good story.'** It's so short and succinct you can tweet it.

And people then say, 'But that's only two things. Have you forgotten one?'

And I say, 'Yes. Most people forget the third thing. That is being clear on what you want to achieve.'

'Is it that simple? Really? So I just need to memorise that and I can put this book back on the shelf and go and watch Netflix?'

And then I have to say, 'No, of course it's not that simple! What were you thinking? If it were that simple you would be able to learn everything you need to know from tweets and not from books! You need to know about messaging and trust and different values and so many things.'

But it is a good starting point to memorise those three points and then probe into them a bit deeper. And yes, that's what we're going to do.

But first a story.

I remember giving a conference talk several years ago to a crowded audience, right after lunch and I was working my way through my data slides on why different people had different attitudes to technologies and it started raining. Really, really hard. You could hear it bouncing off the roof above us.

There is something about the sound of rain, and a full stomach, that just lulls people to sleep (okay, my PowerPoint slides might have been a bit data-rich for this audience of agronomists too). I could see eyelids starting to close and people looking at their phones. So I stopped talking. Then I leant forward – not looking at my notes anymore, and I said, 'Let me tell you all a story'.

At once I had the eyes of everyone in the auditorium. The expectation of what I had to say was suddenly more interesting than their emails or Facebook or even a quick snooze. I don't know if I fulfilled their expectations with my story – but the lesson was never lost on me. There is a certain mesmerising power in breaking into a story.

And think of all the different ways we can engage people through stories of some kind – be it a blog post, a live experiment, a demonstration, a TV show or podcast. Most people get that – at least in theory.

The second point, knowing your audience can be a little bit harder.

One of the Ancient Greek Oracles of Delphi (a sort of ancient Google who could speak wisdom after breathing intoxicating vapours that rose from chasms in the earth), stated, 'Know yourself …'. Even though she then tripped off on a vapour high and never finished the sentence, it became the Oracles' motto and was inscribed into the forecourt of the Temple of Apollo at Delphi. It has since been interpreted in many ways, including *know yourself and you shall know the gods, know yourself and know your enemies* or *know yourself and you shall know others.*

But the fact is – which the Oracle was probably trying to say before she got too high on the ethylene or ethane vapours – that if you know yourself, you really only know yourself. And that is a fundamental principle in communicating science to others.

Yes, you know what you like, but most people are entirely different to you. They are different in what they think and what they feel and what they prefer and what they don't prefer and what they think about science – regardless of what you think and feel and prefer.

We are a very diverse lot, but we love hanging about in similar thinking and acting tribes, which sometimes blinds us to the perspectives of others.

Of course, there are many people out there who understand that science helps us make sense of the world around us, through discoveries based on analysing data, measuring impacts and drawing conclusions from evidence-based thinking. But there are also many people out there who see science as a rather limited way of explaining the world, and who think that it is elitist, blinkered and hampered by its insistence on obtainable data.

There are many other competing world views and perspectives and preferences out there as well, and different types of people will respond best to different stories. That can make getting just the right story to communicate science to an unknown audience quite tricky. But I'm sure you know that already.

There are also many different and competing views as to what good science communication is and isn't. Rather than settle on any one method or model or theory to do my impression of the Oracle of Delphi with, I prefer to think of them all as pieces of a larger jigsaw. While each is competing to be the vital bit you hold in your hand to place into the puzzle, none of the individual pieces can see how they relate to the others until that puzzle is actually completed.

That is your job as a science communicator, to find the bits of the puzzle that you need to put together, to form the bits of the picture you need. You probably don't have to see the whole picture, just the bits that are relevant to you in your work or research or whatever it is you do. But you do need to find those pieces.

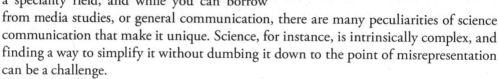

What makes good science communication?

Define 'good' in a subjective-neutral manner that supports consistent metrics.

The intrinsic complexity of science

Good science communication is something of a speciality field, and while you can borrow from media studies, or general communication, there are many peculiarities of science communication that make it unique. Science, for instance, is intrinsically complex, and finding a way to simplify it without dumbing it down to the point of misrepresentation can be a challenge.

Another story.

John Steinbeck, the Nobel Prize-winning US author, took part in a marine science specimen-collecting expedition over six weeks in 1940, and wrote about his experiences in the book *Sea of Cortez: A Leisurely Journal of Travel and Research*. He also made an observation on the blockers of good science communication:

> It is usually found that only the little stuffy men object to what is called 'popularization', by which they mean writing with a clarity understandable to one not familiar with the tricks and codes of the cult.[1]

A decade later, clinical psychologist Anne Roe wrote in *The Making of a Scientist:*

> Nothing in science has any value to society if it is not communicated, and scientists are beginning to learn their social obligations.[2]

These observations were written in 1941 and 1953, and here we are, many decades later, and both their statements still stand true. Admittedly those dull men (and women) are getting rarer – though they are far from extinct. Many of the blockers of good science communication can be institutional, and educating science communicators, without educating their bosses, diminishes good outcomes.

The field of science communication also has trouble defining where it sits in the landscape. Sometimes it gets a PR label, with the negative connotations that come with that, and sometimes it is placed too far down the chain of command to have much influence or impact.

But sometimes – well a little more than sometimes – it rises to a particularly difficult challenge, finds a way to connect and brings some of the wonder and fascination of science to an audience that nods its head and gets it. *Really,* gets it. We need to learn from those instances, and examine why they work and apply those learnings.

Alan Alda (yes *that* Alan Alda, of TV and movie fame) is a great advocate of science communication, and he says that the core of all good communication is developing empathy and learning to recognise what another person is thinking.[3]

What shape the future?

In predicting what the new issues for science communication are going to be, leading science communication researchers Matthew Nisbet and Dietram Scheufele wrote a paper entitled 'What's next for science communication? Promising directions and lingering distractions'. They stated that good science communication is going to be based on several things, including better research and better understanding of our audiences.[4]

My favourite two short quotes from the paper are that '**data should trump intuition**', and that '**effective science communication is not a guessing game, it is a science**'.

They point out that the blockers of good science communication can be some science communicators themselves (or the institutions that they work for). Those that just don't get the change that is occurring to the field, based on good research. Those that don't know how to tell a good and relevant story. Those that don't understand their audience.

Or those that inform without investigating.

Or those that educate without engaging.

Or those that …

… well, some other thing to complete the rule of three pattern, preferably with keywords that start with the same letter for strong effect. If you are clever enough to be reading this book, you are undoubtedly clever enough to put this powerful narrative technique into practice and make up the third line yourself.

Anyway, Nisbet and Scheufele also emphasised the need for more science communication initiatives that span a diversity of media platforms and audiences, and that 'facilitate conversations with the public that recognize, respect, and incorporate differences in knowledge, values, perspectives, and goals'.[4]

To understand that is to understand that there is a new way of communicating science that is based on research rather than intuition (my offering to make up the third point in the rule of three).

What to do with what you now know

You need to see if you can answer a few very important questions. First ask yourself why you are seeking to communicate and mull a bit on your answer. You need to be clear on why you are doing this thing called science communication.

Then ask yourself if you clearly know who your audience is, and important things about them like what types of information they prefer and who they trust.

Once you have those clear, see if you can distil your key message into a few points that will resonate well with that audience.

Then look for a good story to be the medium to take your message to your audience.

Key summary points

- Know your key message and tell it in a good story.

- Know your audience and know what you want to achieve in reaching them.

- Knowing what you like is not the same as knowing what others like.

- Science communication can be intrinsically difficult.

- Best practice science communication is based on research and data, just like good science.

2

Why we need to communicate science better

'If you're scientifically literate the world looks very different to you.'

– Neil deGrasse Tyson, astrophysicist and science communicator

The fact you're reading this book means there is a strong chance that you are already convinced we need to communicate science better – and only need to have that belief affirmed. (We'll talk more about that powerful drive to have your beliefs affirmed in Chapter 17.)

If you were given this book and told to read it, however, here are some compelling arguments by people like Albert Einstein and Neil deGrasse Tyson that might make the case better than I could:

1. Science is important for our lives whether we believe it is or not.
2. Sometimes science is not done for reasons, or in ways, we might prefer.
3. Understanding science and why it is done helps us understand science and its purpose.
4. Scientists are increasingly seeing the benefits of engaging with the public with their science and getting feedback on directions it should go.
5. Being a bit more scientifically literate helps us make better decisions about science, which is much better than making bad decisions about science.

Okay, confession time, they were my compelling arguments, not Neil deGrasse Tyson's nor Albert Einstein's (but I'm channelling them, yeah?). That was an example of 'framing', to make you feel you can trust the statements more. We'll talk more about framing in Chapter 12.

The real answer to why we need to communicate science better is a lot more complex than you can summarise in a few bullet points though. It includes the history of our relationships with science, the complexity of science and society, who owns the tools of science, and who are the winners and losers when it comes to benefits and risks, and it is about the different ways that different people view science and who should be making decisions about it. And that's just for starters.

We can try and demonstrate how complex things can be made a little easier through an analogy.

Consider **Professor X**. He, or she, has been a scientist for many decades and firmly believes that the public's role in science is to pay their taxes so he or she can just get on with his or her research.

Then there is **Professor Y**, who believes that everybody would love his or her research if only they understood what he or she was doing.

And **Professor Z**, who believes that everybody should be as excited about his or her science as he or she is, and if only they understood that excitement, they'd become huge fans and support his or her work in many ways.

And **Professor A**, who is working on a contentious science, say synthetic biology, and knows that it will be important to gain social licence through public approval to attract funding for a continued ability to do his or her research.

And **Ms B**, who is a real science junky and can't get enough of it, and who just wishes that there were more good science communication around.

And **Mr C**, who doesn't really get science, and feels threatened and isolated by it.

And **Miss D**, who likes science, but feels it is controlled by the rich and powerful, and is worried about that.

And **Mr E**, who doesn't trust science and has his own alternative beliefs about how the world works.

And **Mrs F**, who is too busy with her kids and her work and paying the mortgage to even think much about science.

So many different perspectives of science and its communication. All of which have to be considered as valid perspectives that need to be taken into account.

Given all these viewpoints, what is a person to do?

Well for starters we could surely hook up Professor Z with Ms B, and let them get on with their mutual enthusiasm for science. But you'd then have to ask, what about the rest of the population? When science becomes contentious – like in cases of climate change, evolution, embryonic stem cells, infant vaccination, genetically modified foods, etc. – it isn't usually due to concerns from Ms B and her friends. It is more likely from Mr E, who doesn't trust it, Miss D who worries about who controls it, and Mr C who feels threatened by it.

As science communicators we really need to do more than just hook up the Professor Zs and Ms Bs. We need to have strategies to reach all kinds of different people, not just those that are the easiest to reach. And there is a good chance the strategies being used to reach the people we are familiar with might not work very well on those we are less familiar with.

The authors of the tome *The Oxford Handbook of the Science of Science Communication*, have stated:

> Ironically, those communicating about science often rely on intuition rather than scientific inquiry not only to ascertain what effective messaging looks like but also to determine how to engage different audiences about emerging technologies and get science's voices heard.[1]

Professor Z gives a planetary ring to Ms B at their wedding, after having been introduced by a science communicator

This is a problem because communicating science can be derailed by so many things, and intuition cannot take all those factors into account very well. Science communication runs up against misinformation, ignorance, purposeful denial, political expedience and so many other problems. Think about the number of interest groups waging wars on science that result in governments making dubious policy decisions based on bad science – or actively distrusting their own science agencies and scientific advisers.

The alternative?

But we need to communicate about science in difficult times because the alternative of not communicating about science can already be seen. The environment we are now working in has become increasingly complex – with diminished trust in institutions, increasing use of new media, increased rise of 'fake news' and so on. These all lead to what I call the post-truth, post-trust, post-expert world we now all live in (see Fig 2.1).

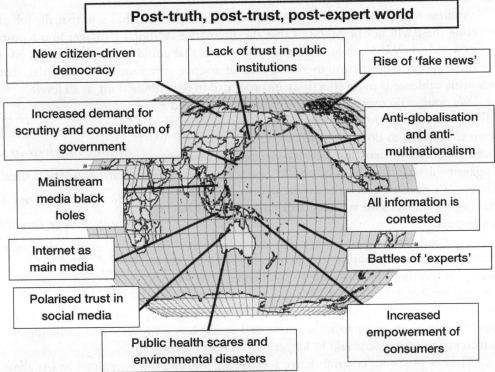

Fig. 2.1. The post-truth, post-trust, post-expert world is complex and constantly changing

Look at African countries suffering drought who banned imports of US food aid because it was from genetically modified (GM) grain and they had been told it was not safe. Or local councils that have banned fluoridation of water for fear it is a serious toxin. Or those parent groups who refuse to vaccinate their children because they don't trust the science behind it.

Individually people have a right to refuse these things, of course, but when you're making a decision that effects a larger population it is not good enough to base your decision on personal beliefs or ideologies. Andrew Maynard, the Director of the Risk Innovation Laboratory at Arizona State University, and an accomplished science communicator, has said:

> What happens when decisions are based on misleading or blatantly wrong information? The answer is quite simple – our airplanes would be less safe, our medical treatments less effective, our economy less competitive globally, and on and on.[2]

The US National Academies of Sciences, Engineering and Medicine has a rather impressive report entitled *Communicating Science Effectively: A Research Agenda*.[3] The report states that some of the key challenges facing effective science communication include the inherent complexity of science information, people's existing biases and attitudes, and a rapidly changing media landscape.

Andrew Maynard was one of the authors of the report. He has said that the job of tackling these will not be easy, but that the alternative is 'slipping further into a post-truth world where disdain for evidence creates risks that could be avoided'.[2] That gives us little option but to turn to the science of science communication in order that scientific evidence is more effectively incorporated in decision making at all levels.

We need to do more than simply explode things and get cheers. We need to do more than measure bums on seats at events. We need to do more than talk to the converted. We need to do more than just talk about the need for more engagement and increased scientific literacy being good things. **We need to know that what we are doing is really having an impact and we need to know how it is having an impact.**

Anyone involved in science communication needs to know that we need do better to have impact in the fraught times we are living in.

The good news is that over the past few decades there has been considerable scientific research undertaken on how science is communicated to different audiences. And when I say considerable, I mean hundreds and hundreds of published academic papers from psychologists and social psychologists and statisticians and media analysts and many other social scientists and researchers on diverse and niche topics – which just to catalogue would be longer than this entire book.

The role of science communicator is changing too, and can sometimes be less about being a translator of the science to the public, and more about being a knowledge broker,

What to do with what you now know

Evidence shows that the two best ways to get better at communicating science are to learn some more about it, and then to get out and do it.[4] But it helps if you are very clear in your own mind about why you want to communicate and who you want to communicate to, as being able to answer those questions in some detail helps define the shape of your communication activities.

There is a lot of very earnest science communication taking place that could be enhanced by a better understanding of the research into science communication, taking it from being a rather blunt tool to a much sharper tool that achieves better results.

So read on, and then go and practice what you learn.

or a match-maker, as more and more scientists become more and more capable at doing the communication themselves.

So to repeat the question, what is a person to do?

Well in lieu of reading hundreds of research papers, you could read a single book like this that tries to summarise the best of them, and tries to give a plain English explanation of what is known, and how you can use what we know to better communicate science.

Key summary points

- The public is very diverse and will not all respond to the same message.

- Communicating science is important, because not communicating science leads to bad decisions, often based on misinformation, biases or pseudoscience.

- The two best ways to get better at communicating science are to learn some more about it and put that knowledge into practice.

3

Burying the deficit model

'There are certain themes of which the interest is all-absorbing, but which are too entirely horrible …'

– Edgar Allan Poe, *The Premature Burial*

Before we get too far into things, we have to bury something called the deficit model. And that needs a background story.

Back in the 18th and 19th centuries, when people had a fear of dying of diseases rather than the vaccines for those diseases, another common fear was being buried alive. It was described well in Edgar Allan Poe's short story *The Premature Burial*, in which the narrator has an obsessive fear of being buried alive. After cataloguing terrible incidents of this having happened to people, the story ends with the narrator waking up in the darkness in terror, believing he himself has been buried alive – but it turns out he was only sleeping snugly inside his sleeping berth on a small boat.

Some people at this time went to great lengths to ensure that if they were buried alive, they could signal their predicament and be saved, and this led to the design of safety coffins. Most had a bell rigged up so that the person inside the coffin could pull a cord, and the ringing bell would notify somebody to come and dig them up. Hence the expression, saved by the bell.

True fact.

Duke Ferdinand Brunswick, in what is now Germany, was said to have had a window installed in his coffin to allow light in and an air tube to provide fresh air. Instead of having the coffin lid nailed down, he had a lock fitted and the keys buried with him. He died in 1792, and there is no record of him using any of these features to get out of his coffin.

Another enterprising German, Dr Adolf Gutsmuth, designed a safety coffin that he demonstrated by having himself buried in, several times. There is no record of him returning after being buried that last time when he finally died.

A side note of interest: one of the drawbacks of many of the safety coffins was that as the buried body decayed it would often swell and shift, causing tension on the cord inside and causing the bell to ring. It caused quite a fright to the poor family members or graveyard workers who dug up the coffin. Clearly a case of a scientific false positive.

Anyway, the point of these stories is that unlike the inventors of most safety coffins, the deficit model just refuses to stay buried, and no matter how many times it is proclaimed in need of burial, somebody imagines they hear that bell and it just keeps being dug up again.

Why deficit model?

The term deficit model was first coined back in the 1980s by an English social scientist, Brian Wynne, who achieved near rock star status among science in society scholars. You could read a couple of dozen scientific papers on it, but in short, the deficit model goes like this:

> The deficit approach implies that there is a lack of accurate public knowledge about science and that improvements in public knowledge of science should increase attitudes, such as public support for science and positive evaluations of scientists.[1]

Put simply, the model implies that people make 'wrong' decisions or have 'wrong' attitudes to science simply because they don't have the right information. And if only the right information was given to them, they would think more positively about the science under consideration. They have a deficit of correct information.

But the model is wrong.

People have their own ideas and beliefs and knowledge and are not 'empty vessels' waiting to be filled with science information. The US National Academies report, *Communicating Science Effectively*, dismisses the deficit model as:

> ... particularly insufficient when people may need to decide whether to take an action and what action to take. The model assumes that if an audience fails to act in a manner that some consider to be consistent with the scientific evidence, either the communication needs to be better crafted or delivered, or the audience is at fault for not knowing enough about the science or not being sufficiently appreciative. ... however, people do not make decisions based solely on scientific information, but take values and other considerations into account.[2]

And the deficit model particularly does not work just because you feel it should work.

The deficit model does not work for so many reasons. It does not work because it is *not* a lack of information that causes people to be concerned about science and technology. And it does not work because providing alternative information rarely changes a concerned opinion. And it does not work because we all think our own information is worth more than somebody else's information. And it does not work because information rarely outranks feelings. And it does not work because if people have a big concern about science, they are not particularly likely to trust a scientist or science communicator.

You can test this next time you have a heated family discussion with your parents or partner, or kids, or colleagues or anyone, on a topic they feel strongly about. Give them your version of facts and just see if it changes their mind. Try and convince them to change their vote or their sporting team allegiance, or to dump their dud boyfriend or girlfriend, or to stop using Apple and use an IBM computer.

But you've probably learned the hard way how that most likely turns out.

Yet the idea behind the deficit model makes a lot of intuitive sense to lots of people. It's as easy a belief to fall into as the narrator of Poe's story's belief he has been buried alive. The only trouble is that like many of our intuitions, it isn't supported by evidence. I know a lot of new age online gurus might extol trusting your intuition, and everybody will have a story or two of when it proved right, but let's not forget all those times it let us down:

- like trusting that really nice salesperson
- like just knowing this time you were going to hit a jackpot
- like thinking a person on Tinder would be a really good match for you
- like believing that car was going to run really well and never let you down
- like (insert random intuitive belief) when in fact (insert disastrous outcome).

So in short, when it comes to communicating science, be a little wary of your instincts if you don't have any valid scientific data to back up what you are doing. **And definitely don't extrapolate a good communication methodology and message from a sample size of one (when it is yourself!) to attempt to reach a wider and diverse audience.**

Because chances are you will have dug up the deficit model and are parading it around like a stinking infectious zombie. Dan Kahan from Yale University (another sci-comms rock star), who has undertaken significant work on how our cultural biases impact our thinking, and who will get a few mentions later in this book, has put it a little more strongly:

> Not only do too many science communicators ignore evidence about what does and doesn't work. Way way too many also shoot from the hip in a completely fact-free, imagination-run-wild way in formulating communication strategies.
>
> If they don't rely entirely on their own personal experience mixed with introspection, they simply reach into the grab bag of decision science mechanisms (it's vast), picking and choosing, mixing and matching, and in the end presenting what is really just an elaborate just-so story on what the 'problem' is and how to 'solve' it.
>
> That's not science. It's pseudo-science.[3]

What to do with what you now know

Recognising the deficit model can be one of those things that you more easily identify in other people's communication work than you can ever identify in your own. But if you start with the premise that most science communication is guilty of using the deficit model until proven innocent, it can urge you to look a bit closer at what you are doing and more critically examine whether your activities are really based around trying to get your audience to think a bit more like you (rather than letting them think like they think).

Strive to become very critically honest with things you are working on, and if you find even a hint of deficit model – there is probably quite a bit of it there – hit it on the head, dig a pit and bury it.

And don't neglect running just as tough a critical eye over your next effort too!

Key summary points

- Really wanting a deficit model approach to science communication to work is not enough to make it work.

- Don't presume you will change people's minds with more information.

- Don't trust your instincts without checking them with evidence.

- Don't extrapolate on a sample size of one (particularly when it is yourself) to prove anything (particularly when you are trying to prove it to yourself).

4

Objective! Your honour!

'When in doubt … C4.'

– Jamie Hyneman, *Mythbusters*

'Let's build a website. We really need a website. Wait! Even better, let's make a video. We can put it on YouTube and watch it go viral. Yes, that's what we need. No. Wait! Let's blow something up. Everybody loves it when you blow something up, right?'

Who has a boss like that? (Or who *is* a boss like that?)

These approaches describe two all-too-common types of science communication:

1. Shoot from the hip communication
2. The School of BSU (blow shit up) communication.

And while they aren't quite as bad as deficit model thinking (which is well and truly buried now, yes?) they still aren't best practice. Why? Because any good science communication should start with a clear objective. Thankfully management scholars have spent many years analysing objectives in laboratories and in the wild, and have come up with some good formulas for explaining their behaviours.

We all know a successful management formula needs a catchy acronym to give it credibility, like CLEAR, HARD, DUMB, ACTION and QUEST. One of my favourites is SMART, coined by George T Doran in 1981.[1] The acronym stands for:

Specific
Measurable
Achievable
Relevant
Timely

The real secret of setting an objective is being able to then measure if you achieved it (which admittedly can be quite hard sometimes). So let's test a few scenarios.

Objective 1: Make a YouTube video that goes viral.
Specific. Well the goal is very specific about what it is that you want to do, but not so much about *why* you want to do it. What exactly is the video going to be of? What do you hope to actually achieve by making the video? Will it be to educate people? Will it be to engage them in debate? Will it simply be to raise general awareness about

something? Of all the SMART letters, if you don't get the S right, you are setting yourself up for a fail. So let's just say the video is going to be of a scientist talking about nanotechnology, and explaining why the world isn't going to be taken over by grey goo (the theory of self-replicating nanotechnology-sized machines getting out of control and consuming all biomass on earth).

Measurable. YouTube has great analytics to measure your views, which is a plus, but just saying 'to go viral' isn't quite specific enough. You should always be able to attach numbers and metrics to your goal. And what exactly constitutes viral? YouTube personality Kevin Nalty (aka Nalts) has said that the benchmark used to be a million views, but as of 2011 that was raised to more than five million views in a three-day period, and that has only gone up since then. Others say you need at least 10 million views in a day to be considered viral. So just pick a very big number.

Achievable. Well that's a bit of a problem as there is just no guarantee that your video, no matter how well produced, is going to go viral, so I think it fair to say this part of the objective is a bit over-ambitious.

Relevant. This can be another tricky one. You really need to show that your communication activity is going to be doing something that aligns with the needs of your organisation as well as your audience. If you are aiming to dispel concerns about grey goo, I think you'd first need to demonstrate that concerns about grey goo are actually an issue for people. You might also be responding to something that is in the media-sphere already.

Timely. Be specific about the time your video is going to take to be produced and go live.

So an improved objective would be:

Objective 1: To make a YouTube video explaining our institute's work on nanotechnology that responds to recent online stories about the fears of grey goo, that is published in one month and reaches 50 000 views in its first six weeks.

Okay. That's a pass mark. Now let's look at another.

Objective 2: Blow some stuff up.

Specific. Not a lot of specifics in this one to be honest. Blow what up? Do it where? In front of what audience? And what do you want to get from the audience apart from a big 'wow'? So let's say the purpose is to explain some basic principles of physics that can be harnessed in rockets or internal combustion engines.

Measureable. Just getting an explosion to happen isn't going to cut it here. You want to measure something more realistic, like the number of people who come to witness your demonstration of basic principles of physics. Or even better, measure how many learn something (yeah, I know that's not always easy to actually do).

Achievable. Do you have the required safety equipment and maybe the right permissions to do this? Do you have the right space to do this? Can you guarantee audience safety? Have you an experienced blower-upper on hand?

Relevant. I think you might need to explain to me once more exactly why you want to blow something up, and what the audience is going to learn from it. And think very hard before you use words like 'engagement with science' or 'scientific literacy' in the same sentence as 'blow stuff up'.

Timely. I'm sure you get this one now, so let's just say 10, 9, 8, 7, 6, 5 …

An improved objective would be:

Objective 2: Demonstrate the inherent explosive power within common household products to 120 school children in four sessions inside a controlled laboratory, to teach some basic laws of physics, to support the school curriculum.

Now that's an objective that works for me to the point that I'd even give you the safety gloves and glasses and matches – though I'd still want a guarantee that you don't get too carried away and forget to link the science to the boom.

Of course, objectives can change as operating circumstances change. For instance the Hague-based Organisation for the Prohibition of Chemical Weapons (OPCW) has had a general communication objective of raising awareness about the Chemical Weapons Convention and the work the organisation does in relation to it. But after the high-profile use of chemical weapons in 2018 in the Syrian War, the unsuccessful nerve agent assassination attempt on Russian double-agent Sergei Skripal, in Salsbury in the UK, and the successful assassination in Malaysia of the half-brother of North Korean leader Kim Jong-un, there was sudden huge media attention on the OPCW's work – but also significant misinformation circulating. So the objective suddenly changed to focussing on increased factual reporting.

When it comes to defining the purpose of your science communication activity, that rather useful report by the US National Academies of Sciences, Engineering and Medicine, *Communicating Science Effectively,* lists five broad goals. The report states that each of the goals places quite different demands on the knowledge and skills of science communicators and their audiences, going from rather simple to more complex.

They are:

1. To share the findings and excitement of science.
2. To increase appreciation for science as a useful way to understand and navigate the modern world.
3. To increase knowledge and understanding of science related to a specific issue that requires a decision.
4. To influence people's opinions, behaviour, and policy preferences when the weight of evidence clearly shows that some choices have consequences for public health, public safety, or some other societal concern.

Graduating head of class in the school of BSU

What to do with what you now know

Have a look at your science communication activities and see if you can articulate a real clear and measurable objective. One that isn't just big picture motherhood goodness – one that is really specific. You are doing science communication in general for big picture motherhood (parenthood?) goodness but this activity you are doing now has a very specific and very measurable reason why you are doing it.

If you can't articulate the objective easily, however, then you know what to do. Go back to start, do not pass go, do not collect your Monopoly money, and begin over until you can articulate that clear objective.

Having an objective keeps you on task, makes you focus on what you are doing and not get distracted too much by the how, and also gives you something to measure. For more on how to evaluate your activities and the types of tools you can use, see Chapter 16.

5. To engage with diverse groups so their perspectives about science (particularly on contentious issues) can be considered in seeking solutions to societal problems that affect everyone.[2]

You may have noticed that none of these said, 'To get a big wow!'

Key summary points

- Before you get too excited about a science communication activity, make sure you have a clear objective so you can measure if you achieved something.

- A useful guide for objective setting is SMART (Specific, Measurable, Achievable, Relevant, Timely).

5

What do the public *really* think about science?

'We live in a society exquisitely dependent on science and technology, in which hardly anyone knows anything about science and technology.'

– Carl Sagan, pioneer

I love standing up at conferences in front of an audience of scientists or science communicators and asking them what percentage of the population do they think just don't get science? Some of them shake their heads like I'm going to be describing a near-mythical creature, or something long extinct like the dodo. I mean, who doesn't get science, right? Others guess it might be about 15%. Or maybe even as high as 20%. Maybe.

Then I make sure everyone is sitting down and flash up a slide on the screen that shows 40% of the population are not really interested in science.

That makes people shake their heads a bit, like learning the Easter Bunny isn't real or that rich corporations don't pay their fair share of tax. Because it means that at least four out of 10 people standing in line at the supermarket are not thinking of the wonder of the Higgs Boson particle, or the contrast between the nanoscale and the expanding universe, or even about how cool it would be to have Neil deGrasse Tyson as a Facebook friend. (Frightening, I know.)

Then they might start questioning the data – like good scientists. And it's true, depending how you ask the question, you can get very different answers. If, for instance, you ask people in the supermarket queue if they think science is important, they will most likely say 'Yes'. The same way they will say 'Yes' if you ask them if cheese is important. Or poodles. Or digital watches.

But if you ask them about how often they have a conversation about science at home or at work, or how often they watch a science show on TV, or read about science in the newspaper, then numbers tend to drop a bit. They drop even more if you start asking them to compare the importance of science in their life to things like staying employed, enjoying good health, making sure their kids get educated and fed, and so on. Even though science underpins many of these issues, they are not likely to be top of mind for many people.

Researchers in many different countries have asked survey questions about people's interest in science in many different ways and have gotten some similar and some

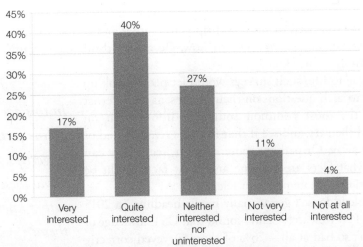

Fig. 5.1. How interested in science are people in Australia? Results from a survey carried out by CSIRO in 2014.[1]

different answers. The 40% uninterested figure comes from a survey conducted in 2014 by the Commonwealth Scientific and Industrial Research Organisation (CSIRO), the Australian Government's leading research institution, shown in Fig. 5.1.

However, another study, carried out by the Centre for the Public Awareness of Science at the Australian National University in 2017, found that more people said they were very interested in scientific discoveries (60%) than said they were very interested in music (38%), films (30%) or sports news (19%).[2] Given, scientific discoveries is just one subset of science, it still looks pretty good, considering Australia is a supposedly sports-crazed nation!

A third study, conducted in 2013 by the Australian Government Department of Industry, Innovation, Science, Research and Tertiary Education, came up with a figure of about 80% believing that science was so important to our lives that we should all take an interest in it.[3] And a fourth study, conducted by Swinburne University in Melbourne, found that the statement 'Science and technology are continuously improving our quality of life' received a rating of 7.24 out of 10.[4]

So clearly, different polls can give different answers, which can be attributed to things such as the methodology, how the questions were asked, and even what questions are asked around them. And honestly, all such surveys should be considered indicative rather than definitive – the same way as if you took a temperature and wind reading at a particular moment on one day, it is only indicative of the temperature and wind speed for the whole day.

We can look at a variety of different surveys conducted in many countries and find a variety of answers to similar questions on attitudes to science. In the US, for instance, figures vary between 87% and 59%.[5,6] Or we could go for the studies that tend to get the media headlines, like only just over 7 in 10 Americans and 6 in 10 Australian know that it takes one year for the earth to go around the sun.[7,8]

The usual response to such survey questions is to lament rising levels of scientific illiteracy. So if you need to weep and rip your clothes and throw dirt on your head, go ahead and get it out of your system now. Because I think such figures are really a distraction from science literacy.

To be true to life, such surveys would let people look up the answers to such questions on their phones, as the internet is cited as the most common source of information on science,[9] and then try and see if that fact was remembered or lost straight away. Or perhaps more relevant, see if they can determine what facts are genuine and what facts might be misinformation on statements on science.

In this age of instant online facts at our fingertips, do we need to know these things, or rather know where to find them?

And while it doesn't get as many media headlines, a 2015 study by the Pew Research Center found that US knowledge of science wasn't so bad at all – 76% of those surveyed correctly knew that ocean tides are created by the gravitational pull of the moon and 73% could distinguish a definition of astrology from astronomy. Compared to international rankings the US far outpolled India, Japan and Russia on most questions.[7,10]

Of course, scientists might think it is a crisis of dumbness that only 35% answered correctly that the property of a sound wave that determines loudness is the amplitude,[7] or only 54% knew that antibiotics will not kill viruses as well as bacteria.[11] The same way legal professionals think it a crisis of dumbness that the public doesn't understand the difference in case law and legislative law. Or the same way that IT professionals think it a crisis of dumbness that many people don't know the difference between RAM and ROM.

Science education

Science education is so complex a topic that the zillions of researchers working on it have developed many diverse and different ways of addressing it – some that work against each other.

But I have attended enough workshops and conferences on science education to understand that it is clearly an 'if only' issue. For instance, if only more kids saw the relevance of science. If only science and technology were a part of all arts and humanities and social studies and history courses. If only all kids understood STEM. If only everyone was scientifically literate. If only more kids really understood what a science career involved. If only we had more highly qualified science teachers. If only we valued science more than sports in society. If only teachers were paid more. If only we had science clubs in schools. If only kids thought science was cool. If only we knew what the actual answer to better science education was.

As it is there are many, many people standing around waving single pieces of a jigsaw puzzle, and no one seems to see they are holding just one piece. If we can put more of these pieces together though, we might see what the picture really looks like.

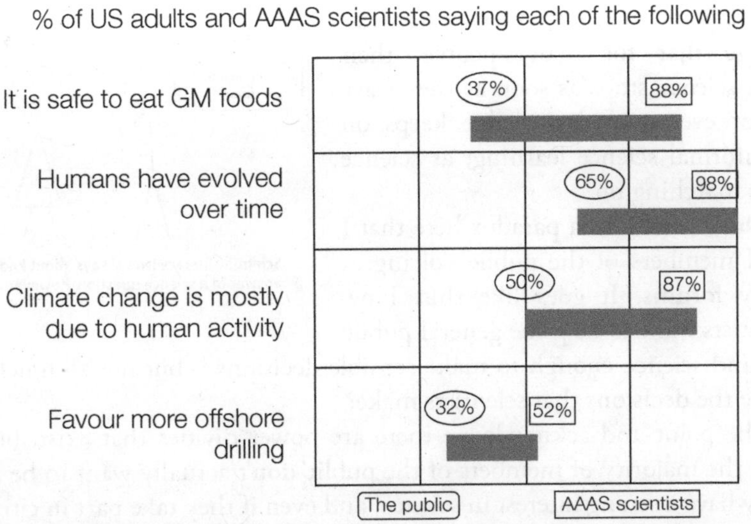

% of US adults and AAAS scientists saying each of the following

Fig. 5.2. Opinion differences between the public and scientists from the American Association for the Advancement of Science, from a Pew Research Center study[12]

So while I could quote survey questions at you all day long, from lots and lots of different countries, there is clearly a trend that not everybody is interested in science, and not everybody knows a lot of basic science (as not everybody is interested in law or knows a lot of basic legal stuff). It should therefore come as no surprise to learn there is a wide gap in scientific knowledge between scientists and the public. Another Pew study, conducted in 2015 and shown in Fig. 5.2, found that while 88% of scientists felt it was safe to eat GM foods, only 37% of the public agreed. A 51-point gap. Or while 98% of scientists polled believed that humans had evolved over time, only 65% of the public agreed.[12]

But that's not the only science-public divide that you need to better understand. It is also important to know that many people think science or technology is advancing too fast to keep up with. And it is important to know if people think the benefits of science outweigh the risks. And it is important to know if people feel the benefits and risks are being shared equitably. **For these can be the things that underpin attitudes to science, and you need to address them if you really want to address attitudes.**

Scientific literacy

It is also important to know if people feel they have enough science understanding to be scientifically literate. However, if we are serious about increasing scientific literacy, we do need to be sure we agree on what scientific literacy means as there is no one generally agreed upon definition. For some it is knowledge of scientific facts; for others it is being able to recite the periodic table; for others it is understanding scientific principles; and for others it is being able to take part in serious discussions about the impact of science on society.

Any discussion of science literacy needs to acknowledge that for many people, their learning of science stops as soon as they leave school. Not everyone's knowledge keeps on through informal science learning, as science itself keeps marching on.

Now there is a bit of a paradox here that I have heard members of the public voicing at community forums. It goes like this: How come scientists want to help the general public

Scientific literacy isn't always about knowing the answer to a science question, however...

to understand science enough to make sensible decisions – but not so much that they could make the decisions that scientists make?

I get the point and acknowledge there are power divides that exist, but I've also found that the majority of members of the public don't actually want to be scientists – even if they have a strong interest in science, and even if they take part in citizen science projects. The same way they'd like to understand tax law but don't want to be a tax lawyer, or understand how to fix their car but not become a car mechanic.

Scientific literacy is a good thing not just because it helps people appreciate the wonder of science, or how its principles are used in our daily life, but because the accelerating pace of scientific and technological advances risks overtaking our capacity to understand it. This used to be true particularly of policymakers, regulators and those in government, but it is now also true for members of the public whose choices and preferences are increasingly taken into account by those people.

Scientific literacy is about understanding the cost and the benefits and the unintended consequences and ethical considerations of science, and then being able to make something like the best decision we can. For in the world we live in, where lobbying and special interests and political ideologies hold sway, we cannot trust that policymakers, regulators and our elected governments always make decisions that we might consider the best they can.

And then there is social literacy

Many issues have a flip side, and for increasing functional science literacy among the public, it is increasing social literacy among scientists. This isn't a reference to that old joke about a shy scientist looking at his shoes when he talks to you, and an extroverted scientist looking at your shoes when he talks to you – this is a reference to the social literacy of understanding what people expect of science, and what methods they employ to gain information or participation in decision making.

It is about having the social knowledge and skills to more fully take part in community discussions in a genuine way. (Okay, maybe it is a little bit of the scientist's shoes joke too!)

What to do with what you now know

If you are serious about increasing the public's knowledge of science, you really need to know what drives their attitudes to science and offer them a lot more than just scientific knowledge. Also offer them the social and political and economic systems that science exists within – and not just how it might impact them, but how they might be able to influence it.

This is easier said than done, of course, but it starts with listening to what your audience knows and what they think, and then being able to provide them with stories of science that fit within many different contexts, such as social, economic and political. For instance, if you are talking GM crops, you may need to talk about the impacts on feeding the world and adapting to a changed climate and the history of crop modifications before you talk about the science that is involved.

There is going to be a lot you don't know about what members of the general public think about science, but a good rule of thumb for science communicators is that they probably don't think about it like you do.

Key summary points

* Different surveys give different findings, but up to 40% of the population only have a small or negligible interest in science.

* Scientific literacy isn't just about understanding science, but also about understanding its social context and being able to participate in sensible decisions about the impacts of science.

6

There is no one public: Making sense of segmentation

'Selling to people who actually want to hear from you is more effective than interrupting strangers who don't.'

– Seth Godin, marketer

Another idea we need to bury is that there is one public. Joe (or Jane) Public is a handy marketing creation to find an average person – but it is one that doesn't actually exist. As the authors of the book *Freakonomics* point out, the average citizen would actually be a person with one breast and one testicle.[1]

Any survey that aggregates the total findings risks misrepresenting the findings. Attitudes to GM foods for instance are often skewed along gender lines, with women being generally more concerned than men. Providing the average finding of a survey inaccurately represents both. That is where segmenting audience findings into distinct groupings provides a more accurate picture. It is also a marketing concept – but a very useful one.

For science communicators, there are many things we can learn from segmentation studies to help us better reach an audience (or audiences), and the first principle to accept is that there is no one public, but lots and lots of smaller publics.

If you think about it, we all come in different shapes and sizes with all kinds of backgrounds and interests and preferences. Some of us like Coke, some of us like Pepsi, some of us like Diet Coke and some of us prefer water. Some of us like fast food, some of us like organic food, some of us couldn't really care what we eat as long as it's there, and some of us consider food so lovingly we take more photos of our meals than we take of our family members.

Likewise, some of us are more interested in astronomy, some of us are more interested in marine biology, and some of us are more interested in the dubious nutritional quality of popular soft drinks. Some of us prefer to read information, some of us prefer to listen to it and some of us prefer to watch it.

We can therefore group people into similar, or like-minded segments. Homogenous sets of a heterogeneous whole, to use the homo and hetero words.

Every one of us is different. But we are also similar to some others.

The trick is to find which segments are the best for you to break your audience into – and bear in mind, we have come a long, long way since Henry Ford said, 'Any customer can have a car painted any colour that he wants so long as it is black'.[2] A walk down any car lot will prove that cars are made for many different tastes and preferences – and come in many colours. This should remind us that you don't choose a communication medium to use because it is the easiest, or the one you prefer, but because it is the one the audience prefers.

Good science communication is not just about changing the thinking of your audience but is also about being willing to change your own thinking.

The makers of cars well know that different types of people tend to prefer different types of cars. Smaller, brightly coloured cars are marketed to young women as louder noisier cars are marketed to young males. Larger families walk right by both on the car lot, looking for SUVs or people-movers. And balding middle-aged guys with more money than sense are hanging around the red sports cars, right?

Well, I'm not actually sure about that. I'd have to go down to the car lot and observe who is buying the sports cars. Which should remind us that stereotyping is a cheap shot and doesn't always conform to reality. In science communication it is very easy to fall victim to common knowledge or presumptions, which may prove to be outdated or may not apply to real life. Rather than make a guess at who is buying those sports cars (though you know that when I say 'sports car' I'm really talking about marine biology or astronomy, right?), go and do some research. Find out if your stereotype fits reality, or if you need to adjust your thinking.

Different types of segments

At the easiest level, people are segmented by gender, age, education, where they live, how much money they earn and other demographics – but we can get a bit more sophisticated than that and also segment them by how they think and feel. This is called psychographics.

After years of studying his patients, pioneering psychologist Carl Gustav Jung developed a matrix of eight key personality types. He believed that all of us experience

Active, passive and hot issue publics

According to the PR guru James Grunig, there are three different publics that you need to take into account on many issues:

- Active publics seek information and enter into a relationship around an issue.
- Passive publics have a low level of involvement and are neither affected by nor see a connection to a particular problem.
- Hot-issue publics arise in response to intense media attention on specific issues such as vaccines.[3]

the world using four key psychological functions – sensation, intuition, feeling and thinking – and one of the four is dominant for a person most of the time. He then added extraverted or introverted personality type against the four traits to get his eight different categories.

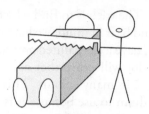

The Myers Briggs Type Indicator Test (aka MBIT) – which has become a bit of a horoscope of modern management – has 16 key personality types, but is based on Jung's work, adding the extra traits of judging or perceiving. It is a form of psychographic segmentation.

The renowned science communicator, Professor Marvello, demonstrates how to segment a target audience

There is an interesting background story here that is too good not to tell. The MBIT was developed by psychologist Katherine Cook Briggs with her daughter Isabel Briggs Myers, after Katherine had met her daughter's future husband, Clarence Myers. Katherine was surprised to find that her views and his were quite at odds and that drove her to research why different people have different perspectives. (There is probably a dumb and blokey joke waiting to be inserted here about mothers-in-law and wives dedicating themselves to analysing a man being the world's longest-running citizen science project, but this is far too serious a book to even dream of making a reference to such a thing.)

So, at the simplest level, you can segment your audience into those who are likely to be interested in your work and those who aren't, and try and find ways to identify and contact those who are interested, and ignore those who are not. But beware of the consequences of only reaching those who are interested, as preaching to the converted is easy, but does not necessarily raise your overall profile or grow your audience.

If you are tempted to get a bit more complex you can look at the diversity of values, attitudes and behaviours across communities, and the different ways they drive how you might communicate with different segments of people.

Internationally there are some very useful segmentation studies you can build on. In the US, the Centre for Climate Change Communication did a study called *Global Warming's Six Americas* (Fig. 6.1), in which they segmented people by attitude to climate change. They called their segments the alarmed, the concerned, the cautious, the disengaged, the doubtful and the dismissive.[4]

A similar CSIRO study conducted in Australia, and shown in Fig. 6.2, found five key segments: the sceptics, the abdicators, the undecided, the eco-friendlies and the eco-warriors.[6]

Once again, you'll see the key to successful segmentation is to come up with cool and catchy names. Though not in every case. In 2014, a New Zealand study came up with the following segmentations for attitudes to science: Penelope Public, Optimistic Oliver, Anxious Angela, Negative Nellie and Bunsen Burner Barry.[7] But they were seen as failing the cool and catchy test and had to be withdrawn.

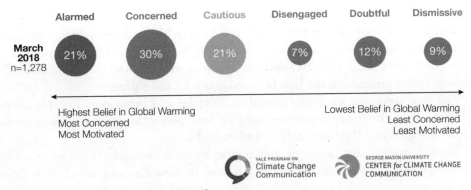

Fig. 6.1.   Global warming's six Americas[5]

Looking back at the Pew studies quoted in the last chapter, they found that respondents answered the questions about knowledge of science differently by race, gender and level of education. For instance, those categorized as White Americans had an average score of 6.1 correct out of 9 questions on science, compared with 4.8 for Hispanics and 4.3 for African-Americans. The study concluded, quite rightly:

> As with gender differences, differences by race and ethnicity could tie to a number of factors, including differences in areas of study at the high school, college and postgraduate levels and other factors.[8]

So finding differences is one thing – but understanding the causes of those differences might be something entirely more complicated, and should also not be based on simple stereotypes without testing them.

Which brings us to values.

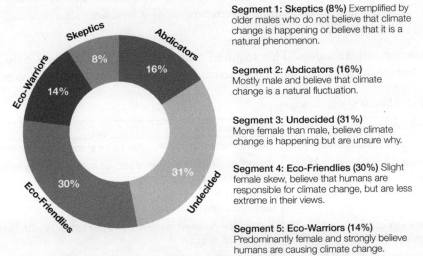

Segment 1: Skeptics (8%) Exemplified by older males who do not believe that climate change is happening or believe that it is a natural phenomenon.

Segment 2: Abdicators (16%) Mostly male and believe that climate change is a natural fluctuation.

Segment 3: Undecided (31%) More female than male, believe climate change is happening but are unsure why.

Segment 4: Eco-Friendlies (30%) Slight female skew, believe that humans are responsible for climate change, but are less extreme in their views.

Segment 5: Eco-Warriors (14%) Predominantly female and strongly believe humans are causing climate change.

Fig. 6.2.   Australian segments by attitudes to climate change[6]

Values

Segmenting gets really interesting when you look at doing it by values, and that is because values underpin so many of our attitudes. We'll get to that in a lot more detail in Chapter 15, but in 2014 CSIRO undertook two segmentation studies of the Australian public to better understand the link between attitudes and values.

The first segmentation study looked at attitudes and behaviours towards science, and came up with six key segments based on how people sought out information on science and how well they were able to understand it.

Similar studies that segment the public by their attitudes to science and or technology have been conducted in the UK[9] and in New Zealand[7] which also showed that different segments have notably different scores for attitudes to, trust in, and understanding of science.

You are probably very familiar with segments 1 and 2 in the Table 6.1, or Mr and Mrs Average, and Fan boys and fan girls, as they tend to be the bulk of those who watch science programs, read science blogs, take part in science public events and so on (and together make up about 50% of the population). They are the people you can most easily reach with science communication activities, and for whom most science communication activities are designed.

Segment 3 can be considered the low-hanging fruit – those you can reach if you just make your stuff a little easier to understand. But then come the last three segments, who together make up around 40% of the public. They comprise people who are either not engaged or uninterested in science and they also don't much value it, understand it, or see the point in it. The CSIRO report says:

> It is no understatement to say that not only is science largely unknown to them, but they are largely unknown to much of science. The people from these last three segments tend to be more likely younger, more likely female, less well educated, more likely to think government funding for science should be cut, and that science is out of control.[10]

Now we turn to values. This required a little bit more work. The CSIRO study asked two sets of 10 values statements across a 10-point scale of agree or disagree. The

Table 6.1. Attitude to science segments[10]

Attitudinal segments	Characteristic
1. Mr and Mrs Average (23%)	Passive interest in science, may come along to science events.
2. Fan boys and fan girls (23%)	Actively interested in science and go out of their way to find science communication events.
3. I wish I could understand this (8%)	Interested in science, but they find they have trouble understanding it when they find it.
4. Too many other issues of concern (23%)	Not really interested in science and just too busy with life to think much about it.
5. Science is a turn-off (14%)	Not interested in science and don't much trust it.
6. I know all I need to know already (2%)	Not interested in science because they feel they know everything they need to know already.

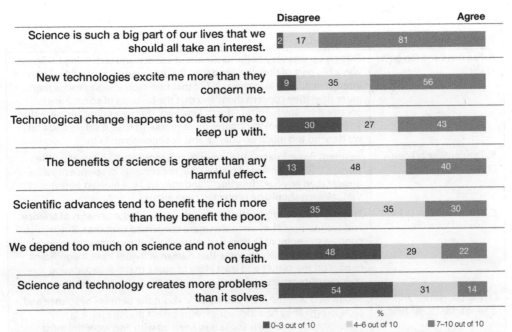

Fig. 6.3. Values statements to science and technology. Attitudes were asked across a 10-fold scale and are broken into the top, middle and bottom thirds of responses for easy of interpreting.[11]

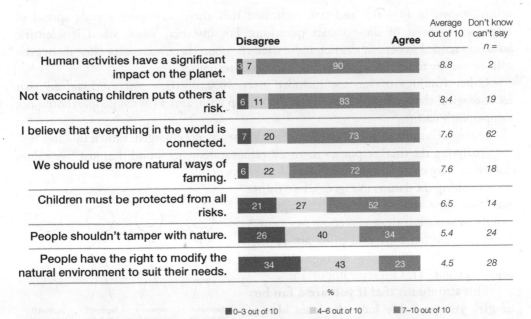

Fig. 6.4. Values statements to the world around us. Attitudes were asked across a 10-fold scale and are broken into the top, middle and bottom thirds of responses for easy of interpreting.[11]

Table 6.2.   Segments of the population[10]

Values segments	Characteristic
A. Fan boys and fan girls (23%)	This group was very enthusiastic about science and technology. Science was a big part of their lives and they thought everyone should take an interest in it. They also strongly felt that new technologies excite them more than they concern them and that the benefits of science are greater than any harmful effects. However, there was disagreement that science and technology creates more problems than it solves and that we depend too much on science and not enough on faith.
B. The cautiously keen (28%)	Segment B was interested in science and technology, but was a little wary of it. They tended to believe that the benefits of science must be greater than any harmful effects, and they had the highest agreement that children should be protected from all risks.
C. The risk averse (23%)	This segment tended to be less positive towards the benefits of science and technology. They were also more concerned with risks. But in contrast to Segment D, they had relatively high awareness of science. They were least likely to agree that human activities have a significant impact on the planet and least likely to agree that not vaccinating children puts others at risk.
D. The concerned and disengaged (20%)	Segment D was the least enthusiastic about the benefits of science and technology. They had the highest agreement that the pace of technological change is too fast to keep up with and were the most likely to agree that science and technology creates more problems than it solves. They also were most likely to agree that scientific advances tend to benefit the rich more than the poor, and that we rely too much on science and not enough on faith.

result, shown in Figs 6.3 and 6.4, indicated that there was quite a wide spread of responses to most of these values questions. For instance, when asked if scientific advances tended to benefit the rich more than the poor, there was pretty close to an even split across the top, middle and bottom thirds of responses.

Other findings of interest included over 40% of those surveyed agreed that technological change happened too fast to keep up with, and 34% felt people should not tamper with nature.

Doing a cluster analysis of the data, four key segments emerged, shown in Table 6.2.

Analysing the total responses to all 20 values statements found that just five values questions were enough to define your values segment.

Graphing the data as in Fig. 6.5, two things stand out: first, the gaps between the segments can differ significantly, and second, Segment A (those science fan boys and girls) are further away from the community average than any other segment. This is pretty important.

This also means that **if you are a fan boy or girl, you probably have the least idea of what might appeal to the other segments.**

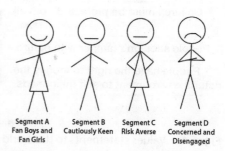

Segment A
Fan Boys and
Fan Girls

Segment B
Cautiously Keen

Segment C
Risk Averse

Segment D
Concerned and
Disengaged

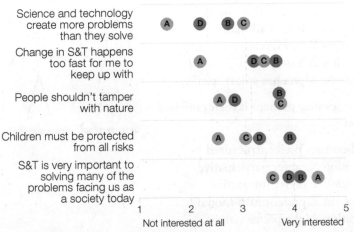

Fig. 6.5. How the different values-based segments map against different values statements[12]

You undoubtedly know what switches on your buttons, but you are probably only guessing what will work for the other segments.

If you are from Segment A, not only will you have quite a bit of difficulty in understanding the values and perspectives of the average community member – but they will have quite a bit of difficulty understanding yours.

And to ask that key question again, so what is a person to do? Well you could start by recruiting members of the other segments into your planning team to help design things that will appeal to them. Or you could simply test things out with your target segment group, if they are not the same as you.

Don't forget X, Y and Z

Differences in the ways that people of different ages think and respond to information is very important in the world of marketing, which science communicators could well learn from. There is a lot of talk about the need to better engage with younger people, but not a lot about the different ways they think that should drive such engagements.

There are many ways of defining the different generations, and one definition, although perhaps a bit too broad to apply to everyone, is:

- A baby-boomer parent would tell a Generation Y child, 'You can do anything!'
- Generation X parents tend to tell their children, 'Do what you're good at!'
- Generation Y's catch cry was 'I want to be discovered!'
- Generation Z says, 'I want to work hard.' But they don't necessarily want to work for other people.

Another breakdown, by types of music devices, which demonstrates their relationships with technology, goes:

- Baby boomers had audio cassettes
- Generation X had Walkmans
- Generation Y had iPods
- Generation Z has Spotify
- Generation Alpha has smart speakers.

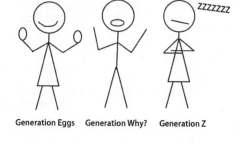

Generation Eggs   Generation Why?   Generation Z

Of more relevance perhaps is their preferred learning style:

- Baby boomers prefer structured
- Generation X prefer participative
- Generation Y prefer interactive
- Generation Z prefer multi-modal
- Generation Alpha prefer virtual.

Generation Z are the first humans in history to be born into a truly digital world. They are currently nearly all in school, and make up 1 in 5 of our population. While only 1 in 10 are in the workforce, in the next decade they will represent over 25% of the workforce.

And after them comes Generation Alpha. The oldest of them are currently still in primary school, but they will be:

- the most formally educated generation ever
- the most technology supplied generation ever
- globally the wealthiest generation ever.

Also, physical maturity is on-setting earlier so adolescence for them will begin earlier. As will social, psychological, educational and commercial sophistication – which can have negative as well as positive consequences.[13]

What to do with what you now know

Repeat three times: There is no one public – there are many publics.

And repeat it every time somebody tells you that you the objective of a communication activity is to reach the general public. Then show them the many ways that the public can be segmented. And particularly show this to yourself.

Being able to segment your audience, whether by age or gender or educational background – or by attitudes and values – helps you hone your messages so that they will best work with those different groups.

You have only one key message that works for different target segments, but it is more likely you might need to refine that message a bit to better work for different groups. After all, the message you develop for school children is not going to be the same message you develop for farmers, or small business, or parents, so why would you develop one message for those who are liberal or conservative, or interested in science versus uninterested?

Key summary points

- There is no one public. We are all different. But we are also somewhat similar to some other people.

- Clustering people into similar groups is segmenting, and is a powerful way to best reach different groups of people.

- You can segment people by demographics, attitudes, behaviours, values or many other measures.

- You more easily understand how to best reach people who are in a segment you are in, but might not understand how to best reach people who are in segments you are not in.

COMMUNICATION TOOLS

7

Messages and metaphors

What the Brigadier sent to the Colonel: 'Enemy advancing from the left flank. Send reinforcements.'
What was received: 'Enemy advancing with ham-shanks. Send three and fourpence!'

– The Temperance Caterer, 1914

You have probably heard the saying by Albert Einstein that if you want to simplify your science in a way that the general public will understand it, you need to be able to explain it to your grandmother. That makes sense (unless your grandmother has a PhD) but what Einstein actually said was 'all physical theories, their mathematical expressions apart, ought to lend themselves to so simple a description that even a child could understand them'.[1]

Don't believe everything you read on the internet just because there is a picture of me with a quote next to it.

Famous quotes from Albert Einstein

I agree, and rather than chatting up random grandmothers I prefer to try things out on children. After all, kids don't have PhDs and they are infinitely curious – but they do need fairly simple concepts, visual explanations, and metaphors and analogies to help them understand complex ideas.

So think of how you might explain your key messages to young kids. Use those types of language and visual explanations and you'll probably do okay. For instance, if you are talking about the relative size of the earth and the moon, it's logical if you are talking to a kid to say, 'If you think of the earth as a basketball, then the moon would be a tennis ball, and it sits about seven metres/yards away'.

And you should be thinking that is the way to describe things to anybody. Because, to be totally frank about this, if you haven't got your message right, all the best science communication theory and practice and support is probably going to be wasted.

To get your message right you really need to be thinking what is going to work best for your audience – rather than what is going to work best for you.

The US National Academies of Sciences, Engineering and Medicine report, *Communicating Science Effectively*, makes a couple of key points to bear in mind about messages:

- Science communicators at a minimum need to be aware that messages from science may be heard differently by different groups and that certain communication channels, messengers, or messages are likely to be effective for communicating science with some groups but not with others.
- Tailoring scientific messages for different audiences is one approach to avoiding a direct challenge to strongly held beliefs while still offering accurate information. People tend to be more open-minded about information presented in a way that appears to be consistent with their values.
- People associate a message with a source, and come to believe the message based on trust in that source. Yet over time people will remember the content of the message more than they remember the source of it.
- It can be difficult to communicate accurate information about science in the face of many competing messages and sources of information.
- It is important to test individual messages carefully before they are used.[2]

The importance of testing your messages is iterated by many reports into developing effective messages, to best address the chances of your message being misinterpreted. This could be due to many factors, including the messenger, audience characteristics or things that interfere with your message being conveyed accurately.

Communication researchers agree that the message a communicator intends to convey is never exactly the same message that the recipient receives.[3]

In general, audiences are more likely to pay attention to messages that are succinct, timely and relevant to them. So whenever possible, find a way to frame a message that includes your audience.

For instance, if you are talking about the relative size of and distance between the moon and the earth, you could say: 'When you look up at the moon in the night sky and wonder how far away it is, you probably think of the many pictures you have seen of it in text books, where it is relatively close to the earth. This is, however, inaccurate. The distance to the moon is about 30 times the diameter of the earth.'

Use of language

I'm sure you have all heard of the KISS approach to language use – Keep It Simple Stupid. It is something to remember as your choice of words can affect an audience's perceptions and responses to your messages. (Indeed, even adding the word stupid to the end of that acronym can insult some people). The no-no list to keep a sharp eye out for includes:

- jargon that has a meaning to you, but maybe not to a general audience
- acronyms, that like jargon, feel like a secret code that the audience is being excluded from
- quotes from academic papers – these often work in the context of a published paper but don't translate too well to messages for general audiences or the media

(the academic way of writing is useful for academic publishing – but almost no other situation at all)

- euphemisms that work for you, but might not be familiar to your audience, like NIMBY (not in my back yard) – both a euphemism and an acronym (as well as a bit insulting) so a triple no-no.
- taking an analogy too far. Analogies are very useful for explaining complex things – but you do need to acknowledge when an analogy breaks down.

The US FDA Report *Communicating Risks and Benefits: An Evidence-Based User's Guide,* cautions that any message you develop based on your own intuition should set off alarm bells. The report states:

> It's exciting to have a flash of insight into how to get across a message. Indeed, the best communications typically do start out as an idea in the communicator's head. Unfortunately that's how the worst ones start out, too. Such intuitions can miss the mark, in part, because the communicator is often not part of the target audience.[3]

Getting the message right

Good messaging, as opposed to the no-nos, is based on:

- having pretested your key messages with the audience
- shaping the message to the particular needs of the target audience, so it is:
  - ➤ relevant to audience needs
  - ➤ sensitive to audience situation
  - ➤ accessible in format, language and channel
- giving the key messages at the start of the communication
- telling the target audience who you are and why you are talking to them (if speaking to them)
- using pictures and stories for illustrating things
- checking your audience has not only received your message, but understands it
- concentrating on addressing not just what you did, but the 'So what?' or the 'Why does it matter?' of your message
- repeating your key message in different ways.

The message box

Washington-based science communicator Aaron Huertas says there are lots of tips, methods, schools of thought and best practices for developing effective messages in science communication. And he likes them all.

By that, he means he doesn't stick to just one method, and views each as a tool rather than a rigid formula.

Having your key messages thought out and then written down can be a good way to focus your thinking. I've been in workplaces where the key messages of the organisation

Case study: Testing a communication message

The US National Academies of Sciences, Engineering and Medicine had been concerned that uptake of engineering among young people was being hampered by it being perceived to be too much about maths skills. Of course, maths is important for engineering, but it is not the only thing you need to succeed as an engineer, and the Academy wanted to find other messages that might resonate with young people better.

So they devised a study that set out to:

- identify a small number of messages likely to improve the public understanding of engineering
- test the effectiveness of these messages in a variety of target audiences
- disseminate the results of the message testing to be used by the engineering community.

The study found, among other things that different messages worked better with different genders. For example, boys found 'Engineering makes a world of difference' and 'Engineers are creative problem solvers' both very appealing messages. But while girls found 'Engineering makes a world of difference' also very appealing, the second most appealing message for them was 'Engineering is essential to our health, happiness, and safety'.

The study found that this optimistic, inspirational statement emphasised connections between engineering and ideas and possibilities, rather than just looked at math and science-based ways of solving problems.[4]

are printed out and mounted on the wall, so nobody has an excuse for forgetting them when they are suddenly asked to tell what the agency is all about.

Having a simple message can be difficult for anyone trained in science, as you have spent years trying to provide context and more data and more precision – and now you have to abandon that for a simple cut-through message.

Aaron Huertas advocates using a 'message box' – or a simple template for writing down your key messages. He suggests they should look like this:

- Here's what we know.
- Here's what's new.
- Here's why it matters.

Or put another way:

- The basic science.
- The new finding.
- The implications for scientists or society.[5]

This template is adapted from the Union of Concerned Scientist's *A Scientist's Guide to Talking with the Media* and an example of it in use is:

Basic science: The Caribbean has the 2nd highest rate of HIV in the world and it's the leading cause of death for people aged 20–59.

Finding: Tourism areas combine several risk factors, including drug use, sex work and population mixing.

Implications: This finding can inform where policymakers target HIV interventions.

Not too bad, but not as punchy as it could be. That's when you look at your messages or test them on somebody and find how well they work. Aaron Huertas then offers version two that is a bit more engaging:

Basic science: The Caribbean has twice as many people living with HIV/AIDS than California.

Finding: We know HIV thrives where drugs and sex work and people come together – and that is in tourist areas.

Implications: Policymakers should recognise that tourist hot spots are HIV hot spots too.

Another message template is the COMPASS message box, which looks at the problems, benefits, so what and solutions to an issue[6] (see Fig. 7.1).

We could fill it in addressing science communication like this:

Issue: Coming up with a simple key message about your science can be difficult when the science is complex.

Problems: Simplifying the science does not do it justice and ignores much of the nuance.

Benefits: A simple message has much greater chance of being understood and accepted.

So what: If science is not communicated effectively it is more likely to be misunderstood.

Solutions: Find tools to use that help you better develop key messages about your science.

You might notice that there isn't a section there for methods – and that's really because no one really cares too much about your methods. Sorry everyone, but it's true. Keep your methods out of your key messages unless they are *really* relevant.

A good rule of thumb for key messages is to have three and no more than five. And odd numbers work better than even ones. We have some sort of program in our head to

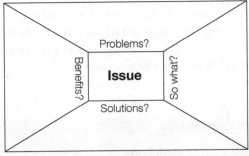

Fig. 7.1. Message box[6]

like things in threes. (Blame fairy tales if you wish – three little pigs, three billy goats gruff, Goldilocks and the three bears– but I think they are tapping into the rule of three too).

Aaron Huertas also says that it is not easy to find a good and simple message that effectively reaches an audience:

> But that's no excuse to throw our hands up and bemoan how hard it is to come up with messages that work. It just means we need to be more creative, more adaptable and more in tune with how people consume information. It also means being aware of our own biases; the messages that feel right for our fellow nerds can carry cultural baggage that make them duds or that backfire with other audiences.[5]

For instance, for those working in environmental sciences or climate sciences – and several other fields – there might be a temptation to tell it like it is and bemoan the uncertain and bleak future that awaits us. The trouble is, that can actually disempower anyone from doing anything about it and feeling it is all too hard. Better, often, if you talk about actions that people can do that might make a difference.

And getting a good message doesn't have to mean dumbing down the science. Alan Alda, who went from TV and movie-stardom to establishing the Alan Alda Centre for Communicating Science at Stony Brook University in New York, has said you can communicate the excitement of science without dumbing it down.[7] But you do need to have a thorough understanding of what you are talking about if you want to develop accurate metaphors and analogies to better explain your science.

Or, as Albert Einstein is also often attributed as saying, with no evidence that he actually did: 'If you can't explain it simply, you don't understand it well enough.'

I never metaphor I didn't like

There are two important things to know about metaphors. First, metaphors are a special type of meta-force. Second, metaphors matter.

Metaphors can make complex science seem simple, or they can capture the nature of something without long and detailed explanations. Science teacher and film-maker Alom Shaha, in an essay on metaphors, begins with some good examples:

> When you think of an atom, do you picture a mini-solar system, even though you know this is wrong? Did you go to a school where the biology teacher made you draw pictures of policemen, security guards or even an army to represent the immune system? Or perhaps, like me, you had a chemistry teacher who told you that positive ions bond to negative ions like lovers in an ardent embrace who, if separated, would rush back together with astonishing speed.[8]

Other popular metaphors include:

- The cosmos is like a string symphony.
- Genes are selfish.

- Nature loads the gun – nurture pulls the trigger.
- There is an endless battle between thermodynamics and gravity.[9]

Though let's be honest, we can get a bit distracted with metaphors of war for battling diseases, or engineering metaphors to describe the mechanics of cells, and computer metaphors for the brain. The problem of such heavy use of these metaphors is that they can dominate thinking and make it difficult for other researchers or science communicators who might have more accurate metaphors.

There has actually been an ongoing debate waged in letters pages and blogs over the good, the bad and the ugly uses of metaphors in science, triggered largely by a blog post in *Nature* in 2013, entitled 'Communication: Mind the metaphor', that looked at the amount of computing and engineering metaphors being used in the biological sciences.[10]

It is true that you should choose a metaphor carefully, as your attempt to explain the science might cause confusion. The following tips are a useful user's guide to finding a good metaphor.

- Carefully consider the idea you want to communicate and your audience.[11]
- Involve experts from different fields (including your target audience) to ensure that language is crafted and used in a way that the goals of effective science communication are met.
- Avoid making value judgments and instead use metaphors to help your audience to understand scientific findings, rather than just trying to convince them of something.
- Make sure your metaphors are culturally appropriate (e.g. a Thanksgiving metaphor won't work well with somebody who did not grow up in the US).
- Craft metaphors in a way that their message is not disproportionate to the evidence at hand.[12]
- A good metaphor should ring true to the science it is describing (albeit that no metaphor is ever perfect, which should not stop you trying to find a useful metaphor).
- It should carry some core truth about the science to someone else.[13]

Science writer Mo Costandi says that metaphors and analogies are probably most effective when they make a tricky concept easier to understand. He says he writes almost exclusively about neuroscience, and often uses ants as an analogy for how the brain works.[14]

Some metaphors, however, have been accused of over-simplifying things to the point of spreading inaccurate information. CRISPR/Cas9 is a good example. At the broadest level, it is both being described as 'ground-breaking' and a 'breakthrough' in attempting to convey the newness of it and its precision. But when it starts attracting unwarranted attention by those opposed to gene technologies it is also described as 'nothing new here. Just a continuation of the technology'.

More complex metaphors for CRISPR/Cas9 use editing analogies, such as taking an encyclopaedia and changing a single word within one sentence, thus altering just the

---

Half-life your message

A team of US researchers and practitioners, slaving in their laboratories late into the nights, in the interest of benefiting humanity, have developed a very useful tool for honing your communication messages that they call 'half-life your message'.

The concept is pretty simple. And it only takes three minutes. It works like this:

1. First you tell your message in 60 seconds – with no preparation, just talking off the top of your head.
2. Tell your message again, but this time you only have 30 seconds, forcing you to concentrate on the key elements of the message.
3. Do it once more in 15 seconds.
4. Do it a final time in only 8 seconds. And you can be sure by the final turn you are concentrating on the real key message.

The team says that message prioritisation can be a major challenge for experts, who tend to have a wide amount of expert knowledge that gets in the way of determining what information is key to give to others, but this exercise focuses the mind on key things.

The exercise can be improved by doing it with a colleague who times you and gives you feedback, or recording yourself – but the team says that following the exercise critical reflection is very important. For instance, is the final message the most appropriate one for your audience? Have you left out something crucial?

The exercise can also be improved by having first having a brainstorm on the topic you intend to talk about, in order to 'promote mental access to relevant ideas and creative flexibility'.[17] Or, after running a half-life exercise on that terminology – to get your creative juices flowing.

---

meaning of that sentence and nothing else – are perhaps a little overstated and possibly misleading.

Beast or virus?

To demonstrate the power of metaphors, researchers from Stanford University gave participants a brief passage about crime in a hypothetical city named Addison. A few words were altered in the text of some, so that the passage said that crime was a 'beast preying' on the city. For the others, crime was described as a 'virus infecting' the city.

Simply changing the metaphor influenced people's beliefs about crime. Those given the 'beast' metaphor were more likely to believe that crime should be dealt with by using punitive measures, while those given the 'virus' metaphor were more likely to support reform measures.[15]

Another study by one of the same researchers, with other colleagues, found that the metaphors used to describe global warming can influence people's beliefs and actions. They asked 3000 Americans to read a short online news article about climate change. As in the previous experiment, there were slightly different versions of the story. In one, the metaphor 'war against climate change' was used, and the other the 'race against climate change' was used.

What to do with what you know

Developing a strong and effective message is a key to successful science communication, and while some great messages are developed through intuition, it is generally a good idea to distrust intuition and test messages with your target audience to understand what they are taking away from them.

It is also important to try and maximise the environment you are communicating in to ensure that your message is received well and is best understood. If talking to an audience this involves removing all the distractions that might prevent your message getting across and working to make your message relevant to the audience.

So work hard on your key messages, test them for effect, and then repeat them – in a variety of ways – every chance you get. Having a good message is like having one straw, and you won't break the camel's back (no camels were actually harmed in the development of this metaphor) unless you repeat it enough times in enough ways to build up that critical mass to do so.

Those who read about the 'war' against climate change were more likely to then agree with scientific evidence showing it is real and human-caused, compared to those who read about the 'race' against climate change.

The researchers suggested that war metaphors remind us of other war-related concepts, like death and struggle, which remind us of the negative feelings and consequences of being defeated in war, and the importance of being victorious.[16]

So use the 'meta-force' wisely, young Jedi.

Key summary points

- Messages need to be kept as simple as possible, but also made relevant to your target audience.

- Avoid acronyms, jargon, euphemisms and the type of expressions you'd use in an academic paper.

- Don't just explain what you did, explain the 'So what?' of it.

- Test your messages with your audience to ensure they resonate with them.

- Metaphors are a very effective way of getting across complex information, but they need to be accurate and tested for effect.

8

Once upon a time: Storytelling

'Scientific theories ... begin as imaginative constructions. They begin, if you like, as stories, and the purpose of the critical or rectifying episode in scientific reasoning is precisely to find out whether or not these stories are stories about real life.'

– Sir Peter Medawar, biologist

Once upon a time there was a science communication book that set out to visit grandma's house – sorry, I mean it set out to show that the way you tell a story has a huge impact on how well you manage to communicate with others. After all, storytelling is the way we learn about life as children. It is the way we learn about what is right and what is wrong. It is the way we learn our family histories and the way we learn to relate experiences about life to each other.

So why then, when it comes to telling the stories of science, are we so uniformly bad at it? Is it because the art and structure of storytelling has been beaten out of us through the education systems, leaving us with a strict structural approach to how a science narrative must be told? Is it because we feel that reverting to a story-based structure will somehow diminish the importance of what we have to say? Is it because we become too preoccupied with telling the stories that we want to tell, not the stories that people necessarily want to hear?

Or is it that we have just forgotten how to tell a good story?

Regardless of the cause, it is possible to learn – or re-learn – how to tell a good story and how to engage an audience with your work through different narrative structures, use of good metaphors, and lots of other narrative devices that are all backed up by sound scientific research.

But first, let me tell you a story.

Randy Olson is a former marine biologist who took the brave plunge to leave academia, and armed with confidence and a vision, he went to film school and then on to Hollywood, where he battled villains and monsters and encountered wise sages to guide him until he finally emerged as a wiser and skilled Jedi Knight version of a science communicator. And remember that story, because we will get back to it later.[1]

He has made films on science controversies, such as *Flock of Dodos* and *Sizzle*, and has written on the skills you need to tell a good story. His book, *Houston, We Have a Narrative*, argues that we who have not achieved the status of Jedi Knight science communicator, can benefit from using the Force of existing and proven narrative templates. One of these, he says, came from watching a documentary on the making of the cartoon series *South Park*. He says:

> In the middle of the show there was a scene that changed my life. It was extraordinarily profound, and I believe it can transform the entire world of science. The scene featured Trey Parker talking about his technique for editing the first draft of each show's script.

He described that technique as the 'rule of replacing and's with either but's or therefore'. What that means is that rather than listing what happened and happened next and what happened and what then happened after that, change the AND this happened to THEREFORE this happened, or BUT this happened.

This can be given as a scientific formula, called ABT (And, But, Therefore), which you can write down like this:

_____ and _____, but _____, therefore _____.

He says that many stories can be broken down into this structure. Consider *The Wizard of Oz*. There was once a little girl living on a farm in Kansas AND her life is boring, BUT one day a tornado swept her away to the land of Oz, THEREFORE she had to undertake a journey to find her way home.

For a science story, he says you might use the magic formula along these lines:

> I can tell you that in my laboratory we study physiology AND biochemistry, BUT in recent years we've realized the important questions are at the molecular level, THEREFORE we are now investigating the following molecular questions ...

He even goes on to give us a narrative index whereby we can calculate a narrative score for a piece of writing by dividing the number of buts by the number of ands, and then multiplying it by 100. This narrative index, he argues, can be used to explain why Charles Darwin is considered the father of evolution and Alfred Wallace is pretty much forgotten.

Which brings us to – another story!

Once upon a time, back in the 19th century, Charles Darwin was busily touring the world on the good ship *Beagle,* collecting specimens and making notes and closely observing the variations of life that he saw. Wallace, meanwhile, was spending four years in South America and eight years in the Malay Archipelago, closely observing the variations of life that he saw.

Darwin's voyage on the *Beagle* lasted from 1831 to 1836 and he spent the years afterwards writing up his book, *The Voyage of the Beagle*, and developing a theory about why life came in the varieties and similarities and differences that it did.

Meanwhile, on the Malay Archipelago, Wallace had developed a theory of his own on why life came in the varieties and similarities and differences that it did, and he wrote up a paper on it and sent it to Charles Darwin. Darwin received the paper in 1858 and nearly fell out of his chair in surprise, realising it was the same as his own theory!

Wallace's paper, and an extended abstract of Darwin's work (which would later become the book *On the Origin of Species*) were read out at a meeting of the Linnaean Society in August 1858 – along with a letter that Darwin had written to a colleague the year before on his theory.

So, we had two British naturalists with the same theory, being published about the same time. Who would become the pre-eminent researcher?

We all know it was Darwin. And yes, he was in Britain at the time, and yes, his letter to his colleague proved he had primacy of the idea, but Randy Olson argues, there was a little more to it than that. Using his narrative index, he rates Darwin's work as scoring much, much higher than the work of Wallace, making it much more compelling reading.[1,2]

You may or may not be convinced by that as being the reason for Darwin 1 and Wallace 0, but the learning from the story is sound – if your work is competing with another, the one that has the highest narrative score will invariably win out.

It's not simply a case of typing ABT into your word processor and expecting it to do some magical search and replace for you though. As Kristi McGuire of the Union of Concerned Scientists points out:

> Scientists who want to succeed with Olson's methods will have to not only read and process what he has to say, but also commit to thinking about how to communicate their work more effectively over time. This isn't an add-on to doing good science.[3]

Other narrative structures

The ABT structure is just one you can work with, and if you find it doesn't work intuitively for you there are others that you can turn to. Randy Olson says that most common narrative structure that scientists use is the IMRAD model (pronounced, I'm Rad, maybe).

You might say that you've never heard of that model of writing, but let's break down the acronym into its component parts: introduction, methods, results and discussion. Look familiar now?

Or there is the formula most used by Hollywood, known as Freytag's pyramid, after the 19th century German novelist and playwright, Gustav Freytag (see Fig. 8.1). It is broken down into beginning, rising problem, climax, cooldown, and ending or resolution.

Also, in an analysis of over 1700 works of fiction in English, Andrew Reagan and colleagues from the University of Vermont came up with six key story lines. They are, with examples of the story structures:

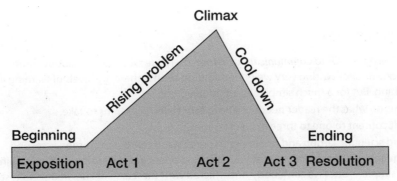

Fig. 8.1. Freytag's pyramid

1. Fall-rise-fall (Oedipus Rex).
2. Rise-fall (Icarus) – many stories from Hans Christian Andersen.
3. Fall-rise (man in a hole) – *The Magic of Oz* stories.
4. Steady fall (riches to rags) – *Romeo and Juliet.*
5. Steady rise (rags to riches) – *Alice's Adventures Underground.*
6. Rise-fall-rise (Cinderella) – *A Christmas Carol.*[4]

The researchers also found that the most popular stories followed the 'fall-rise-fall' and 'rise-fall' arcs, though you might prefer to tell a story with a happier research ending!

Writing plain English

It is often said that scientists need to unlearn how to write in order to write more clearly, but there are some easy guides to help you tell your story in a way that is more likely to be understood. One strategy is to use a readability test on what you have written. In addition to the narrative index mentioned in this chapter, there are around 40 different readability tests that you can choose from. Their effectiveness and ease of use varies.

In short, these tests are formulas to measure how complex your text is. The US Food and Drug Administration, who treat readability very seriously as it can relate to getting drug doses right or wrong, recommend:

- SMOG (the Simplified Measure of Gobbledygook test)
- The Fry Readability Test
- The Flesch Reading Ease Test
- The Lexile Framework for Reading.[5]

The Flesch-Kincaid Readability Test, an adaptation of the Flesch Reading Ease Test, is incorporated into Microsoft Word's readability software, so is easily accessible. (However, it has been pointed out that it frequently presents falsely low evaluation scores.) You can find the others online at several sites with a quick search.

Writing style tips

In *The Chicago Guide to Communicating Science*, there are two chapters about writing – 'Writing well' and 'Writing very well'[6] – reminding us that there are levels of learning in everything. But for a much simpler list of good advice:

- Focus on what the reader needs to know, especially for actions to take.
- Limit content to one to three main messages.
- Avoid scientific jargon and use easier to understand terms. (Yes, jargon is second language to many scientists that you may have spent years learning, and unlearning it is not simple – but find somebody not afflicted with the 'curse of knowledge' to test things on.)
- Use short sentences. And short words. Like this.
- Use active voice and address the reader personally.
- Use positive rather than negative messages.[5]

Little Red Riding Hood and the Big Bad Wolf discuss narrative theory

The power of narratives

In October 2016, 25 scholars from around the world came together at the University of Florida College of Journalism and Communication, to try and understand what makes one story better than another. They came up with a several recommendations, including:

1. **Tell stories about individuals.** Personalise your story. We care and can empathise about the impacts of something on an individual more than we care about the impacts on thousands of people.

2. **Give your audience two plus two.** Leave space for the audience to draw their own conclusion from the story, and don't lecture them on what it all means. This is particularly relevant if you are challenging someone's beliefs.

3. **True stories also have to feel true.** If an audience feels that a story isn't right, or doesn't align with their own beliefs, they might distrust it. Feeling real is really important.[7]

4. **Paint a picture in the mind of your audience.** Visual language leaves a strong memory, such as Martin Luther King's famous *I have a dream* speech in which he stated, 'One day in Alabama little black boys and girls will be able to join hands with little white boys and girls as sisters and brothers.' The holding hands image is powerful.[8]

Another US researcher, Michael Dahlstrom, has stated that narratives have been shown to:

- increase comprehension
- increase recall
- shorten reading times
- increase interest
- increase engagement, when communicating science to non-expert audiences.[9]

Narrative does this by tapping into the parts of our brains that appear to be predisposed to process information more efficiently when it is presented as a story, especially when that information has social relevance.[10]

As was noted in that National Academies report *Communicating Science Effectively*, we must endeavour to better understand the structures and processes that encourage effective science communication, and determine the best practices for communicating scientific uncertainties, to ascertain the role of narratives in communicating science.[11]

We have much we can learn from the humanities here and the way they have studied stories and how stories work on us.

When confronted with lots and lots of useful lists of principles and pointers that potentially overwhelm you with choice, pick the ones that you feel the most comfortable with and go with them.

When science narratives get lost

Regardless of the narrative structure you choose, there are times when the content lets you down (clearly this is the 'rising problem' part of the chapter's narrative structure). For science this often has to do with a reluctance to admit failure, which is actually a crucial part of many narrative structures. Its impact on the public or the many different publics, however, is that they do not have a realistic understanding of how the scientific process works, and when a failure comes to their attention, they view it as a rare and catastrophic failure, rather than a common thing.

Science can very often be about failures, and new research directions that they set. But the public stories that we tell of science go out of their way to delete any mention of failure. Publication is also almost only about successes. Media releases are almost uniformly about successes. And grant seeking is all about successes.

Even the word 'failure' is avoided and terms like 'null result', 'dead-end', 'lack of outcome' and 'unexpected result' are used. And because we do not often tell the stories of the processes of science, and the failures that come with them, when a scientist says, 'just trust us', there is no foundation for that. Because there is no firm understanding of how those results have come about.

In an opinion piece published in *The New York Times*, the physicist and popular science writer Laurence Krauss lamented another problem: the way that most science stories failed to capture the public's imagination. He cited the example of gravitational waves as something that has failed to generate public excitement, despite scientific excitement.[12]

However, Michael Dahlstrom asks if the problem might be in the way much of science is told? He quotes the English author EM Forster as saying, "'The king died and then the queen died" is a story, but "the king died and then the queen died of grief" is a plot'.[13]

The second statement goes beyond simply stating the facts and gives a causal connection between the events. Scientists, Dahlstrom argues, observe events in the natural world and then attempt to draw connections between them, while we might better communicate if we were less emotionally neutral and explained why things happened. Why the queen died, for instance.

He argues that to succeed as science communicators, **we need to go beyond just making the science facts more accessible to a general audience, and engage with them emotionally as well**.

This can be tricky when seeking to be neutral and unemotional in communicating your science, but an example is to talk about the potential impacts of a science. In the embryonic stem cell debate, while those against the technology were talking of the harm to embryos, those in favour framed the discussion around things like being able to potentially cure your grandmother's Alzheimer's disease. It is a good example of using facts that engage emotionally.

Competing narratives

As much as Luke Skywalker uses the Force for good, there is always, it seems, someone who has gone over to the dark side, using the Force for their own nefarious ends. And narratives have a greater potential to convey biased information, due to their high emotional impact and limited need for citations.[14] So it is no surprise that for many contentious issues there are many strongly competing narratives.

An example of competing narratives is when US President Donald Trump publicly worked up support for the US to pull out of the 2015 Paris Agreement on climate change reduction. The Paris Agreement had set a common goal for mitigating climate change, which he framed as harming American workers, who would lose their jobs as a result of it.

In his narrative, he is the Hero, the climate change agreement is the Villain, and US workers are the Victims.[15]

After he had made this statement, many European leaders responded with their own narratives, in which they framed President Trump as the Villain, the planet and climate

Impacts of good narratives

Good narratives, or framing information in stories, have been shown to:
- increase people's likelihood of remembering information.[17]
- reduce counter-arguing.[18]
- make people feel the experience being described was their own.
- be much more effective than a fact sheet.
- be much more convincing than just data. Really![8]

change scientists and other countries as the Victims, and the rest of the world's unity as the Hero. President Macron of France even lampooned President Trump in a speech, saying, 'Let's make the planet great again'.[16]

Countering narratives with other narratives is an effective strategy, but as Marty Kaplan of the University of Southern California says, 'You know the expression "Don't bring a knife to a gun fight"? I submit, "Don't bring a data set to a food fight"'.[17]

The hero's journey

A final word on that introductory story about Randy Olson overcoming the forces of evil to triumph as a film-maker, and why that might have particularly engaged you (yes, this is the 'resolution' part of the narrative structure of this chapter). There is a very powerful narrative structure called the hero's journey, that is the structure that was used in the first *Star Wars* film, and in *Finding Nemo, The Wizard of Oz, The Matrix* and many, many more. It has a few variations, but there are basically 12 steps that take a central character from a normalised world (the Shire, Tatooine, Neo's workplace, etc.) through a series of stages that sees them falter, learn and then eventually overcome their obstacles or nemesis and return home a changed person:

1. Central hero introduced in the ordinary world.
2. Call to adventure.
3. Refuses the call.
4. Meets the mentor.
5. Crosses the threshold.
6. Tests, allies and enemies.
7. The approach.
8. The ordeal.
9. The reward.
10. The road back.
11. Resurrection.
12. The return home (prize).

What to do with what you now know

So now that we all agree that using narrative structures is a good way to convey information, how do you actually do that? If you are not sure, the ABT model is a good one to begin with, and you can build on that to develop more complex narratives when you are more comfortable and familiar with it.

Once you are aware of different narrative structures you will start being able to recognise them more often. And you will undoubtedly see particular structures that you like and feel fit your work. Break these structures down and adapt them for yourself.

Find the ways to personalise your stories and find your best metaphors. These are tools that will serve you very well in many communication situations and circumstances.

It's a good structure to know because it's a very powerful subconscious structure and it neatly tells the story of most scientists in their careers as well. But it also applies to science communication – be prepared for setbacks, and ordeal, but find your mentors and persist and you will eventually reach the prize (although getting super powers and marrying the prince or princess is not guaranteed). Therefore, this structure can be used with great effect to tell a science story. (Did you see what I did there? ABT!)

Key summary points

- We respond to stories emotionally, and we remember stories better than we remember facts and data.

- There are many narrative structures you can learn to apply to your communications – find the ones that work best for you.

9

Trust me, I'm a scientist

'If we really want the public to trust science, we have to create a scientific system that is worthy of trust.'

– Rose McDermott, political scientist

Let's start with a quick poll. Who do you think people trust the most?
Scientists?
– Sorry. No.
Politicians?
– Hahahahaha!
According to polls conducted in Australia, the UK and the US, nurses top the trust charts. A poll conducted by Roy Morgan in Australia in 2017 found nurses topped the trust charts for the 22nd year in a row as the most trusted profession. They were followed closely by pharmacists, doctors, engineers and school teachers. Bottom of the list were those working in car sales, advertising and real estate agents.[1]

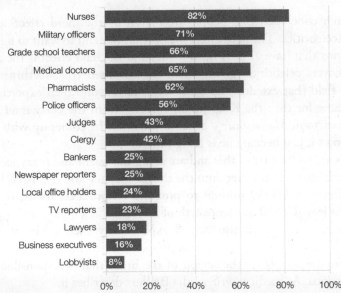

Fig. 9.1. America's most and least trusted professions. The percentage of US adults saying the following have high/very high honesty and ethical standards.[2]

55

A US poll conducted by Gallup in 2018 (see Fig. 9.1) also rated nurses as the most trusted profession. They were followed by military officers and teachers. Medical doctors and pharmacists rated a bit lower.[2]

Also, no scientists! In fact, scientists as such weren't even thought of as being important enough to be included on the list as a profession to ask about!

So who else do people trust? We all know celebrities hold a level of trust way above their qualifications, on all kinds of topics – but just how much are they actually trusted? A study by the Korn Group found that in Australia, the most trusted celebrities are Hugh Jackman, Jamie Oliver and Ellen Degeneres.[3] (At least there were no celebrity health faddists spouting paleo diets and placenta milkshakes on the list.)

In the US, the 2013 Readers Digest poll of trusted personalities listed Tom Hanks, Sandra Bullock and Denzel Washington as the top three most trusted people. The first scientist – Robert J Lefkowitz, MD, Nobel Prize-winning chemist – came in at 11th place, ranking several places behind Alex Trebek, the host of the TV game show *Jeopardy!*[4] Something else to note, TV judges like Judge Judy ranked higher than any judges on the US Supreme Court.

True fact.

But why do people trust celebrities and media personalities so much? Scientists have actually done research on this and a paper published in *The British Medical Journal* looked particularly at shonky celebrity health advice. For instance, Sir Michael Parkinson, former celebrity interviewer, said that you can self-diagnose for prostate cancer by peeing against a wall. According to him, if you can hit a wall from two feet away you don't have prostate cancer. Of course, if you can only hit your own two feet – well then!

The research concluded that people are driven by the 'herd effect' and the 'halo effect' to trust celebrities. The herd effect means that we tend to want to follow the most popular decisions that have already been made, and the halo effect is the angelic beam of fame that covers celebrities, giving them a generalised trustworthiness from their success in one field that extends into other fields far beyond their expertise.[5] (My wife has a better name for this: the King Julian Effect – from the crazy mad leader of the Lemurs in the cartoon *Madagascar* – because whenever he comes up with a crazy idea, everybody follows it just because he is King Julian!)

Advertisers are well aware of this and are using the warm and fuzzy feelings that we have towards celebrities to transfer onto the things they try and sell us. Why else would PepsiCo pay Beyoncé US$50 million to promote its products to us, if we were not susceptible to at least US$50 million worth of influence?[7]

Neuroscience studies have also helped explain why such celebrity endorsements work so well on us. Brain scans shown that images of celebrities increase activity in our medial orbitofrontal cortices – the region of the brain that is responsible for forming positive associations. As medical writer Julia Bellhuz describes it:

> So if you're an Angelina Jolie fan, seeing her image lights up this part of the
> brain, making you more likely to think highly about whatever she is promoting,

even when it's something extreme like a double mastectomy to prevent breast cancer.[6]

There have been no shortage of polls looking at trust from different perspectives. A poll of 11 000 people world-wide looked at which accent people trust the most – and the winner was British! (Okay, the survey *was* conducted by TimeOut online, and asked that question alongside asking which accents were the most sexy, and which cities were the best for dating (Paris and Melbourne BTW).)[7]

And it doesn't stop there. A poll conducted by McClatchy/Marist in the US in 2017 found that most people don't even trust polls. Only 37% of registered voters had any trust in public opinion polling (with only 7% saying that they had a great deal of trust – a figure my wife conjectures might be suspiciously close to the number of people who might work in advertising, marketing and polling!).[8]

Celebrity scientists

But before you think I'm starting to digress a little down a rabbit hole on polling, let's look at celebrity scientists – those rare media stars who are both credible celebrities and also credible scientists.

Declan Fahy, a researcher at Dublin City University, who had previously worked in the US, has looked at the rise of celebrity scientists and how they help the public become more engaged in science. He looked at the careers of Carl Sagan, Stephen J Gould and Neil deGrasse Tyson (sorry David Suzuki, Brian Cox and female celebrity scientists who weren't looked at) as scientists who are both widely known and widely trusted by the general public.

He found that the reason celebrity scientists are widely known, while most earnest scientists never achieve that level of fame, could be put down to a few key factors. First, they needed to achieve professional authority or expertise in the scientific sphere at the start of their careers, and when they moved into the public sphere, they needed to retain the ability and credibility to operate in both. Next, it was important that they never lost their legitimacy in either.[9]

But coming back to the polls, what about those that did look at trust in scientists?

A British poll, conducted by Ipsos-Mori in 2017 found that 83% of the public thought that scientists were trustworthy (though again nurses topped the poll with 91% trust).[10] In 2014, an Australian National University poll found that 71% of respondents trusted scientists.[11] And in 2016, a US survey by the National Science Foundation found more respondents expressed 'a great deal' of confidence in science leaders than in leaders of any other institution except the military[12] (see Fig. 9.2).

The currency of trust

There can be strong and ongoing trust in institutions. In Australia, the CSIRO, the country's premier public research institution, has maintained strong public trust for decades. Such trust can be thought of as a currency that the organisation saves. However, just as the balance in a no interest bank account will diminish relative to inflation, if

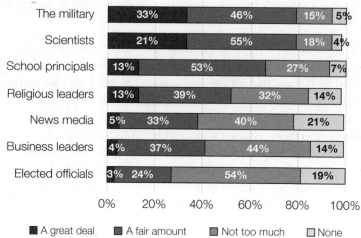

Fig. 9.2. Americans' trust in institutions. The percentage of US adults who say they have confidence in a range of groups to act in the best interests of the public.[12]

you don't use your trust currency, it can slowly diminish too. But every time to do use it – such as making a statement on contentious science issues – you also diminish the bank account of trust (as you would when you spend money from your savings).

So, in a low-trust environment like we now live in, it is very, very hard to add to your trust account, but you have to decide if you want that trust account to diminish from using it or from not using it.

And my advice is to use it, so you are at least seeing some return from it.

The unasked question

Here's a very big question that doesn't seem to have been asked in any polls. If trust in scientists isn't actually too bad, why do many people not much trust science? Several researchers have actually poked at it with a stick to try and find out the answer.

According to the Pew Research Centre in the US, there are distinctions in those who trust scientists in general, and those who trust scientists on contentious issues, such as childhood vaccines, climate change and genetically modified foods. The Pew report stated:

> Overall, many people hold sceptical views of climate scientists and GM food scientists; a larger share express trust in medical scientists, but there, too, many express what survey analysts call a 'soft' positive rather than a strongly positive view.[13]

In the US, many people are either sceptical about scientists' level of understanding of contentious issues, or think scientists disagree on them. On climate change, while 28% of the public feel that climate scientists understand the science behind it very well, 32% feel that they don't understand it. And for GM foods, while 19% felt that scientists

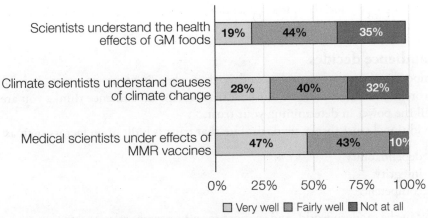

Fig. 9.3. Public confidence in the knowledge of scientists in the US[14]

understood the science very well, 35% felt they did not. And 35% of the public felt that only half or fewer of scientists did not agree that climate change was due to human activities, and 53% felt half or fewer scientists did not agree that GM foods were safe to eat[14] (see Figs 9.3 and 9.4).

Nobody rejects science as a whole, of course. People love their smart phones and their TVs and they are generally happy with the weather forecasts, but they are less happy when science conflicts with a deeply held view, when they are being asked to believe something that goes against what they feel is true. And that can be based on their political beliefs ('My party does not endorse that!'), religious beliefs ('My God did not say that!') or personal beliefs ('That's not how I was raised!').[15]

Or as celebrity social scientist (at least in academic circles if not in the tabloids) Dan Kahan of Yale has put it: 'The issue of climate change isn't about what you know, it's about who you are.' And added to this are overlays of your position on race or ethnicity,

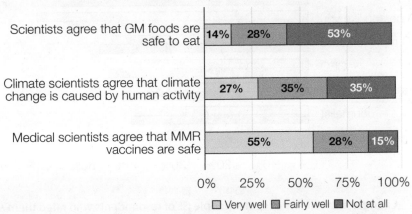

Fig. 9.4. Public belief in scientific agreement on contentious science in the US[14]

income, religiosity, social capital, education, and knowledge – all of which can affect levels of public trust.[16]

The audience decides

An important concept to understand is that when it comes to trust, it is not how trustworthy you think you are, but how trustworthy an audience thinks you are. They hold all the power in determining your trust.

Traditional trust theory tends to state that trust is built upon factors such as:

- dependability
- integrity
- competence.

But in the modern world, and in more recent research, relationships are seen as more important. This reflects both the changing nature of the world and changing ways that trust is developed. Another facet of the modern world is that trust is no longer given, it is only loaned, and will be withdrawn very easily when trust is betrayed.

One of the best overviews of trust across society is the global Edelmann Trust Barometer (see Fig. 9.5). The 2018 edition was produced from online surveys in 28 countries and it looks at paradigms of trust, including looking back at factors of trust over the past 18 years. For instance, in 2001 there was a rise in influence of NGOs. In 2005 trust shifted from authorities to peers. In 2006 a 'person like me' was the most trusted. In 2011 there was a rise of the authority figure. In 2017 there was a growth in inequality of trust. And 2018 has seen the Battle for Truth.[17]

The general finding for 2018 was that there has been an across-the-board collapse in trust, of governments and institutions, with a general feeling that the 'system' is failing people and that concerns are becoming fears.

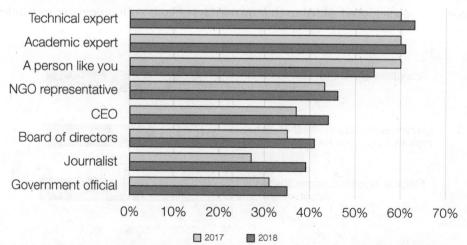

Fig. 9.5. Changes in credibility of spokespeople (% of respondents who rated them as trustworthy) between 2017 and 2018[17]

The study even divided respondents into the two key segments of more informed members of the public versus the mass population, to determine what impact that had on trust. Within these two segments people were divided into the three groups of trusters, neutral and distrusters. It found the informed public were more trusting than the mass public who had a huge trust collapse.

It also found there was general collapse of trust in the media with less than 50% of people consuming news not even once a week. That's right – every second person does not read or watch or listen to news at least once a week! Bear that in mind if your work is based on getting your message out through the media (see Chapter 10).

And if that doesn't scare you enough, they found plenty of examples of the echo chamber of our beliefs in action, with 55% of people not regularly listening to people or organisations with whom they often disagree. **They were also four times more likely to ignore information that supported a position they did not agree with, and had more trust in search engines than human editors when having information selected for them.**

The Edelman report concludes:

> The trust crisis demands a new operating model for organizations by which
> they listen to all stakeholders ... engage in dialogue with them; and tap peers
> ... to lead communications and advocacy efforts.

Or, as Marcia Kean, of Kean Healthcare, has described it, our authoritative trust fabric has deteriorated. She has argued that a new social contract is needed between science and the public that reflects the landscape in which science is conducted today, and the new ways in which the public intersects with and consumes scientific findings.

To address this, she has said, 'A new trust fabric of partnership, participation, and peer groups will need to be built.'[18]

Trust versus distrust

Put simply, people's willingness to act on scientific information has been shown to be influenced by trust.[19] But distrust of science appears to have a bigger impact on the effectiveness of science communication than trust, so countering distrust by building up levels of trust is crucial.[20]

And, perhaps counter-intuitively, distrust is not always the opposite of trust. Nor is it an absence of trust. Distrust is more based around lack of credibility or perceived willingness to deceive.[21] But again, this can be more perception than reality. And as we have discussed already, when you have an adversarial environment, and have two competing messages – one that says, 'Trust me for a complexity of reasons!' and one that says, 'Do not trust you for a simple reason!' – simplicity will always win out.

The implications for professionals and organisations that deal with complex issues is fraught.

Science communicator and physicist, Helen Czerski, author of *Storm in a Teacup: The Physics of Everyday Life*, has said that in the age of Google, the frontiers of knowledge are misleadingly comprehensible rather than inaccessible. This ease of access means we often don't see the complexity of context before arriving at conclusions.[22] So we tend to look for simple answers to complex questions, and it is the simple answers that rarely do justice to the complexity of the issues involved.

And it is rarely the scientists giving simple answers.

Test if for yourself by doing a Google search and finding if it is easier to understand what is online about paleo diets and placenta milkshakes than it is to decipher actual scientific information on health and wellbeing.

But what does it all mean?

The simple answer to that is that trust is hugely important, and efforts need to be spent on both building up trust and stopping it from being eroded.

But how do we do that?

The answer according to many researchers is through better engagement on science. Well, yes – but what does that actually mean? Put simply, it means building an effective relationship between scientists and the public. (Sorry single guys and gals, not *that* type of relationship!)

Researcher David Kipnis has stated that in evolutionary psychology terms, we tend to trust those people who share the same gene pool as us.[23] In the modern world, however, our tribes are a little different, and they tend to be those people who rant and rave like us. Or who are outraged by the same things as us. Or post the same memes on social media as us. Family and friends still rule consistently as the most trusted sources for advice, but after that, when you really want somebody who knows what they are talking about, we tend to look for those people who think like us or share our values and beliefs.

Of course, astute readers will already have picked up the problem here. If you have wacko beliefs (let's say that wearing tinfoil hats stops mind-reading aliens from probing your thoughts), and you want to talk to someone about it, or seek someone's advice on some aspect of the science of how aliens actually do this, then you go looking for someone who has similar beliefs to you.

You might as well be asking the mirror.

And let's not forget the lesson we learned, or *should* have learned, from all those alien invasion films of the 1950s and 1960s, that when I say 'aliens' it is really a metaphor for the government – the way it used to be a metaphor for communists. That's right, you don't actually wear a tinfoil hat to protect yourself from alien mind probes. I mean who believes that? But government mind probes! Now we're talking. (See the strange case of David Icke in Chapter 17.)

Trust me on this.

Next let's look at the reason why so many say that they trust science and scientists, but only on certain topics. There is a school of thought that goes, if there is a lack of trust in science and scientists, it is because kids aren't taught enough science at school, and if only kids knew more science they would understand the position scientists are taking better on all those contentious issues. But you are smart enough to know that is deficit model thinking, right, and that way lies madness and blindness and years of living in a small apartment like a bitter deposed despot ranting at the injustices of the world. Okay, maybe not that extreme, but you get the point, yes?

According to Daniel Willingham, professor of psychology at the University of Virginia, of more benefit is to educate people about why they are prone to accept inaccurate beliefs in the first place.[24] British Baroness Susan Greenfield, professor of pharmacology and physiology at Oxford University, believes lack of trust in scientists is also something scientists have contributed to. She has said that the public has become wary of scientists and are cynical about things like GM food and 'Frankenstein sciences', because they are frustrated that scientists don't really give them clear answers to their questions, such as do some vaccines cause autism or do mobile phones cause cancer?

She says scientists can no longer ignore engaging with the public, and the inability of scientists to come out with simple yes or no answers to complex questions only makes people look for those who can provide the simple answers. She also says that if people better understood how science works, through scientists getting into the popular media more often, there might be less distrust.[25]

So let's look at the facts and put together the equation:

People most trust those they know and those they feel have similar values to them.
Let's call that **TSV.**

+

People trust scientists on many issues, except when they disagree with their findings.
Let's call that **TXDF**

+

Simpler answers to complex issues are more readily understood.
Which we'll call **SA2C**

+

People love celebrities – and often feel they have some relationship with them.
And we'll call that **♥C**

So the equation is: **TSV + TXDF + SA2C + ♥C = FFFV.**

Or to spell that out for the mathematically challenged, science and scientists will be more trusted if people are more Familiar with scientists, if they Feel they have a relationship with them, if scientific answers could be Framed more simply and be more easily understood, and if people understood why their Values distorted their trust (FFFV).

It might not be the full equation and answer, but it's a good start.

Trust me.

What to do with what you now know

We all get it that in times of diminishing trust it is important to both establish and maintain trust if you wish to be seen as credible by your audience. But it is also important to know that trust tends to be established in the first few moments, so first impressions really are vital.

Remember, in many situations it is more important to be seen as empathetic than knowledgeable when seeking to reach your audience (as will be covered in more detail in Chapter 15). Don't start by telling everyone what you know. Start by telling them what you feel – about how glad you are to be talking to them, or how much you love their town/profession/passion/your own interest in coming to talk to them, etc.

Be a person first and a scientist second.

Particularly if you can show you are a person like your audience in any way (a parent, an employee, a farmer, etc.). As we will mention later, people want to know that you care before they care what you know!

Key summary points

- Trust is important to have people pay any attention to what you are saying.

- Traditional models of trust don't work so well any more.

- People most trust friends and family and people who they perceive as having similar values to themselves.

- People trust celebrities more than they trust experts generally, as they are people they like and that transfers to trust.

- Establishing relationships and familiarity with your public can help improve your levels of trust.

10

Media matters

'Both science and journalism are a search for truth.'

– Robyn Williams, science broadcaster

I was at a conference in Wellington in New Zealand a few years ago, in a lecture theatre full of social scientists who were debating the role of the media in society. The conversations went something like this: 'The media is more than a mirror to society as it sets the agendas from which community conversations flow.' 'We need to better challenge the power structures of the media and the way they select and decide what is news.' 'The media selects not just the news agenda but the power imbalances that they reinforce.'

All of a sudden a woman at the back of the lecture theatre stood up and said in a loud voice, 'Listen everyone, I'm a working journalist, and it's pretty clear to me that not one of you has ever spent a day in a newsroom in any media organisation in your life. For if you did, you'd know that agenda setting is really more about the chaos and randomness by which something becomes news or doesn't. There are so many factors from the individual journalist to the sub-editor and the balance of stories for the day that go into deciding what is news– and it is a very chaotic and random environment.'

Well, everyone in the room went pretty quiet after that – but her point was well made. It is one thing to analyse the media at a distance and postulate about how it works and what its power base and agenda is, but it can be quite another to actually work inside a media organisation where the rush and bustle of getting news produced seems to take precedent over any theories or agendas.

The media can be a bit like that famous elephant in the room (not the elephant in the room that people don't want to acknowledge – because that's actually a gorilla in the room, right, that has slowly morphed into an elephant through misuse). The elephant in the room is the one where different people go into a darkened room and touch a bit of the elephant and each think they have a different animal. The one who touches the trunk thinks it is a snake, and the one who touches the leg thinks it is a hippo and the one who touches the tail thinks it is a donkey, and so on. Your perspective of the media – or which bit of the beast you encounter in the darkness – determines how you interpret it. And needless to say, those who view the media from inside a newsroom are often astounded at how the media is perceived from outside the newsroom.

Added to different perspectives, the media landscape is one that can change rapidly, which means that what holds to be true at one moment in time might not be as true very shortly thereafter. So be cautious about adopting media communication advice too heartily without checking if the environment has changed.

For instance, while there is much lamenting that there is general decrease in the number of specialist science journalists in media organisations, it has not necessarily led to a decrease in science stories. Dr Susannah Eliott, the Chief Executive Officer of the Australian Science Media Centre, says that she feels there is actually more science in the media than many people realise, and journalists are very happy to use comments and perspectives from scientists – if they are offered.[1] (Secret code for: be a bit proactive, don't just wait to be contacted for a quote!)

In the good old days

I think we have all heard stories about how the media worked in the good old days. You'd write up a media release that was basically the synopsis of your most recent paper, put in a few line returns, fax it out to the newspaper and radio and TV stations, and they'd all be on your doorstep the next morning to cover your story.

I'm pretty sure it was like that when I started working in science communication – but I remembered a study that found that the more you know about a topic, the more likely you are to have false memories about it[2] (or that's what I remember it found). So it is really quite likely that getting a story into the media was never that simple, and that alternative social networks always existed (even before social media existed) and there was always a mix of design and luck in getting a story in the media.

As there still is today.

So let's start by accepting a few key things about the 'meeja':[3]

- The media landscape *is* complex and can change rapidly.
- You are going to have to engage with the media at some time.
- The better you understand it, the better you will probably do.
- Journalists generally do a pretty good job given their time and resource limitations (but know that they are doing *their* job, not *your* job!).

The benefits of using the media are pretty apparent. It can be an effective way of communicating messages about scientific research to potentially millions of people worldwide, and that can lead to new funding and new collaborations. But this doesn't come easily. You need to be very proactive and work across TV, radio, newspapers, Twitter, YouTube, Instagram and so on.[4]

But of course, there are natural concerns that the time it takes to become an expert at the complexity of media will detract from your own work – unless your work *is* dealing with the media. Then your issue may be dragging reluctant scientists into the media spotlight, trying to drag others who get a little too addicted to it out of the spotlight, and stamping on the files of some others 'Never let near the media!'.

Let's first have a look at some aspects of the media that you need to know about, regardless of your background or day-to-day work. We'll do that by looking at some 'truthiness' about the media – or things that feel that they are true – and see if they stand up to actual data.

Truthiness 1: Who really gets it wrong?

Media outlets are often accused of getting science stories wrong, right? I mean, wrong, yes? However, a UK study of health stories found that much of the exaggeration in the media was actually coming from press releases from research institutes that were based upon information provided by the scientists.[6] Another study found that of 200 randomly selected medical press releases in 2005, 29% were rated as exaggerated and less than half provided appropriate caveats to their claims.[7]

People probably don't use the news media in ways you think they do, and are rarely well enough informed or motivated to weigh up complex and competing ideas and arguments.[5]

Due to the pressure on researchers and their media offices to get stories run, there is a common tendency of overstating results and their implications.

Truthiness 2: The media provides false balance

The media's need to present an unbiased portrayal of a controversial issue, by providing both sides of a story, results in digging up a dissenting scientist and that gives the impression that there might be equal amounts of dissent and consensus. This is a false balance that fails to acknowledge how one point of view might only be held by a very, very small minority – and sometimes by a scientist not even from a discipline directly relevant to the topic being discussed (look at the credentials of many of the anti-climate change scientists for example).

A big problem? Maybe less so that it has been. Many scientists have woken up to the fact that false balance is most easily overcome by simply pointing out in an interview that they represent 95% of the scientists or whatever.

Also, many journalists have become more aware of this too, and are trying to move beyond the trap of false balance.[8]

But of course, it is still exists. If you suspect you are being set up in a debate that supports a false balance:

- insist on expertise
- be clear where the burden of proof sits
- stay focused on the point at issue and not some side track to the debate.[9]

Truthiness 3: The media is more about entertainment than information

Well that's a big one – and it's certainly true that many areas of the media may seem to be more about entertainment than information, catering to celebrity hosts rather than experts, and looking for telegenic pictures over substance. But we need to acknowledge that there is media and there is media and there is media – and you need to be aware of the type of newspaper, magazine, TV show or online journal that you are engaging with and if they are likely to diminish the seriousness of your story or not.

Although, interestingly, a US study by the Pew Research Center found that science news that was packaged as entertainment actually increased participation by those not very interested in science.[10]

Truthiness 4: The media overplay the risks at the expense of the benefits of new science or technologies

As we will look at in Chapter 18, when dealing with risk communication, you need a whole different set of tools than you do for normal communication. And yes, the media love a story about risk because it is intrinsically more newsworthy than a story about benefits. However, that doesn't mean you will always have the risks of your work overplayed at the expense of benefits.

I often tell people that working with the media in a risk communication situation can be a risk in itself – but sometimes it can be a bigger risk not to work with the media. And understanding how the media operate in risk communication situations can empower you a lot to keep control of your story. Some useful things to know include:

- Claims of risk are not only usually more newsworthy than claims of safety, but also a much simpler story.
- Reporters do their jobs with limited expertise and time and may gravitate towards the simplest explanation.
- In risk stories reporters tend to cover viewpoints, not 'truths'.
- A risk story is usually simplified to a dichotomy.
- Reporters prefer to personalise a risk story.[11]

Truthiness 5: Social media has replaced mainstream media as a source of information

Actually not. It does depend a bit on whether you count online versions of media outlets as social media – but let's just think of them as online versions of the mainstream media and put social media into another category. In that instance, television is still the main form of information for the majority of people, regardless of whether they watch it on their phone, on a tablet, on a PC or on a television. A UK Ipsos Mori poll conducted in 2014 found that British people were more likely to get science news from television than other mainstream media. The poll found that of 1749 British adults surveyed (and 315 youths between ages 16 and 24) almost 60% got most of their science news from television. Only 23% said print news was their primary source, and even fewer – just 15% – read it on online newspapers or websites.[12]

In the US, however, a more recent 2017 study by the Pew Research Center found that 54% of Americans regularly get their science news from general media. Social media, while prominent as a general news source, was described as only appearing 'to play a modest role in informing Americans about science'.[13] The study also found that

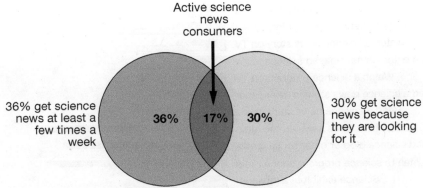

36% get science news at least a few times a week

Active science news consumers

30% get science news because they are looking for it

36% 17% 30%

Fig. 10.1. One in six Americans both actively seek out and consume science news[13]

only 17% of people are active science news consumers, with 36% being described as getting science news at least a few times a week and 30% as getting science news because they are seeking it (see Fig. 10.1).

Most social media users see science-related posts on these platforms, though only a quarter (25%) see 'a lot' or 'some' science posts; and a third (33%) consider this an important way they get science news.

The Pew study also looked at where people get their science news from and whether they felt that source got the facts mostly right. For science-based media the results were consistently above 50%, while those who used a range of news outlets felt that only 28% got it mostly right (see Fig. 10.2).

In Australia, television was also the number one source of science information. A CSIRO study found that in 2014 television was the most popular single medium for getting information on science and technology by the general public. However, those

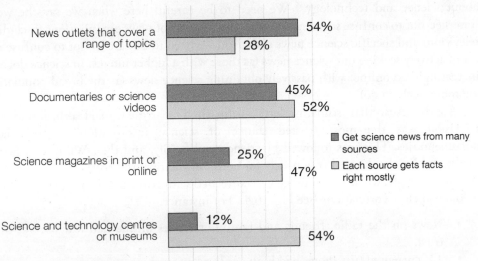

News outlets that cover a range of topics — 54% / 28%

Documentaries or science videos — 45% / 52%

Science magazines in print or online — 25% / 47%

Science and technology centres or museums — 12% / 54%

■ Get science news from many sources
□ Each source gets facts right mostly

Fig. 10.2. Belief in whether sources of information get the facts right in the US[13]

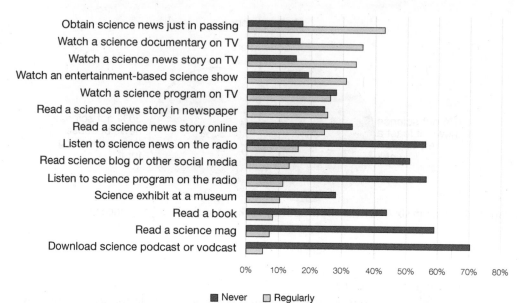

Fig. 10.3. Most popular media for science information in Australia[14]

with a high interest in science favoured the internet. The study found television was the preferred medium for obtaining science news by 32% of the population, with the next closest being news websites at 24%.[14]

However, the way science information was obtained was not in directly seeking it out, but just finding it in passing – which was stated by 43% of respondents[14] (see Fig. 10.3).

This is important to understand in light of findings such as a US National Science Board study in 2016 that found over 45% of Americans turned to the internet for news about science and technology.[15] We need to be careful here when we say the word 'internet' not to confuse social media with the websites of news organisations (including television) and specific science news sites. And we need to be careful not to confuse the act of actively seeking out science news (as those with a higher interest in science do, and increasingly do online) with passively obtaining science news (as the broad population are more likely to do).

Another Australian study, undertaken for the Department of Health, found that television was the most trusted source of science information – particularly documentaries. This was followed by friends and family, and then Wikipedia. Social media and Facebook rated very poorly for both information and trust.[16]

Interestingly, when media usage was correlated with trust, the data showed that few media had close correlations (see Fig. 10.4). For instance:

- News on the radio, friends and family or stories online had low use yet high trust.
- TV current affairs shows had high use but low trust.

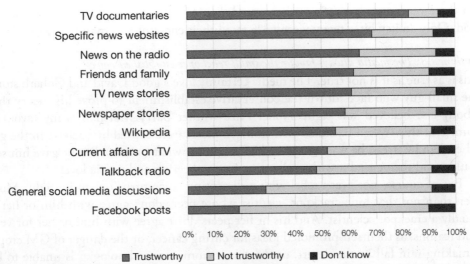

Fig. 10.4. Most trusted sources of information in Australia[16]

- Social media including Facebook, and talkback radio, had high use, but low trust.

Truthiness 6: The media has become an echo chamber for ideologies
Guilty as charged. But honestly, when has it not been? Haven't the establishment media always supported the establishment and radical media supported radical ideas and public media supported communitarian ideologies?

It is much more transparent now though, as we move from any single major source of media (which we probably last saw with the town crier) to now having such a breadth of media to select from that we can easily choose those that best accord with our preferences and ideologies.

Social media has brought with it many impacts – and one of its main impacts on the mainstream media has been to vastly accelerate the news cycle. This leaves less time for analysis and debate, thereby moving it from a relatively trusted source of information into just another filtering machine – competing to give us news that we approve of.[17]

Truthiness 7: Scientists and the media have a difficult relationship
Well you certainly hear this a lot, and you have as much chance of being bailed up by a journalist who has had a few too many drinks complaining about scientists as you have of being bailed up by a scientist (maybe even at the same event) who has had a few too many drinks, complaining about the media and journalists. However, a study of 1354 biomedical researchers in five countries found that they had largely positive interactions with the media.

The study, which was conducted in the US, Japan, Germany, Great Britain and France, found that about 70% of the respondents had interacted with the media in the past three years and that 75% of those rated their encounters with the media as mainly good. Only an itsy-bitsy teeny-weeny 3% rated their encounter as mainly negative.[18]

Truthiness 8: The media always favour the small rebel over established voices

This is as true as it is not true. The media certainly love a good David and Goliath story. The small guy who faces down the conservative establishment to prove his theory that nobody else believed was right! Nobel Prize winner Barry Marshall is my favourite example of this. When he stated that stomach ulcers were caused by bacteria in the gut rather than by stress, he was poo-pooed (if you know what I mean). But he gave himself an ulcer and then cured it with antibiotics. Great cheering and media love!

But of course, not every rebel with a dissenting voice turns out to be right. More often than not that lone crackpot scientist whose peers don't agree with him or her *is* actually a crackpot scientist. And his or her peers don't agree with him or her for very good reasons, as their self-promoted ideas for curing cancer, or the danger of GM crops, or making rain fall in the desert, or the threat from nanotechnologies is unable to be backed by any real evidence.

So yes, the media do love a good David versus Goliath story, and it is pretty easy to get coverage for your story if you are a dissenting voice. It is also fairly easy to rationally and effectively demonstrate why the dissenting voice is not agreed with by the majority of researchers – but you must do it a way that does not make you seem as bullying as Goliath. If you try and censor a lone voice it will look like you are picking on the little guy. Andrew Wakefield's finding on autism being caused by the measles, mumps and rubella vaccine quickly fell into the David versus Goliath narrative and he got a lot more media sympathy at first than his story deserved.

According to the UK Science Media Centre, the way that the science and media communities deal with 'mavericks' is crucial to whether their research is lauded, decried or ignored.[19] So it can be more effective to remove someone's underdog or 'David' status by bringing them into the fold. Then you scrutinise their work as a peer, rather than a maverick.

Truthiness 9: It has become hard to get science stories into the media

Well yes and no. A study of science news content across the three major US TV networks (ABC, CBS and NBC) did find that for the years 2013–14 science news made up only about 2% of nightly news. And there has been a trend in most countries that specialist science journalists are diminishing in numbers across all types of mainstream media. When a science story is reported nowadays, it is often given to a journalist who might have a background in politics or environmentalism or some such, and they will report the story through that frame.[20]

But that means that you are more likely to get your science story covered if you can give it a political, social or other frame that is consistent with what is being covered in the media.

If the theory of evolution has taught us anything, it should be that we need to adapt to changes and anticipate changes in the future.[20]

Truthiness 10: Being popular in the media can be detrimental to your science career

Once upon a time there was certainly evidence to support this. Carl Sagan is often held up as an example of someone who was refused tenure at Harvard University, and was 'blackballed' (if you know what I mean) from becoming a member of the National Academy of Sciences because his media popularity was seen as distracting from the seriousness of his science. However, this doesn't really hold up anymore and most major science agencies and institutions, including the Australian Academy of Science, the British Royal Society, the US National Academy of Sciences and the American Association for the Advancement of Science now actively encourage scientists to engage more with the public and the media.[15]

A theme in truthiness?

Which actually brings me to the point of looking at truthiness statements about the media. Yes, there can be some truth in them – sometimes a lot – but of the ones we have looked at, you actually have the ability to influence how true or not they are. After all, working with the media is a two-way relationship, and you do have some capacity to influence how that relationship plays out.

Four US Researchers – Joshua Conrad Jackson, Ian Mahar, Michael Gaultois and Jaan Altosaar – in an opinion piece in *The Conversation*, have argued that it should be the responsibility of all scientists that better science is getting into the media.[21] This is becoming something more and more scientists and science communicators are advocating.

The researchers argue that if you do a general search for some scientific answer to a question that is fundamental to the existence of our species, like, 'How can I lose weight?', you are much more likely to come up with a long list of recommendations that are not based on science. Some might even refer to 'studies show', without ever stating who did the study.

But accessing good science is difficult, as it is often hidden behind jargon and paywalls. (And, sorry science guys and gals, but if your research is locked away behind paywalls you are actually a part of the problem, not the solution!)

They argue that by developing new ways to disseminate science knowledge it might be possible to help prevent inaccurate and over-hyped stories from gaining traction:

> Because of recent research, we know there's little evidence that genetically modified vegetables are unhealthy, and that eating less meat is a simple way to positively influence the environment. These are important messages, and when people don't hear or listen to them, there can be serious consequences. Misinformed campaigns arise against vaccinations, and near-extinct diseases return.[21]

They state that instead of tolerating media that run outrageous scientific claims, it should be all scientists' personal responsibility to make their research freely available in media that are both accurate and accountable, like *ScienceNews*, *Quora*, *Reddit*, *AskScience* and *The Conversation*.

The Daily Truthiness

Wow! Wacko!
Wizzer!
Whoopee!
Wham!

Wow, wacko, wizzer   Wow, wacko, wizzer   Wow, wacko, wizzer
whoopee and wham!   whoopee and wham!   whoopee and wham!
Wow, wacko, wizzer,   Wow, wacko, wizzer,   Wow, wacko, wizzer,
whoopee and wham!   whoopee and wham!   whoopee and wham!
Wow, wacko, wizzer,   Wow, wacko, wizzer,   Wow, wacko, wizzer,
whoopee and wham!   whoopee and wham!   whoopee and wham!
Wow, wacko, wizzer,   Wow, wacko, wizzer,   Wow, wacko, wizzer,
whoopee and wham!   whoopee and wham!   whoopee and wham!
Wow, wacko, wizzer,   Wow, wacko, wizzer,   Wow, wacko, wizzer,
whoopee and wham!   whoopee and wham!   whoopee and wham!

The next challenge, of course, would be to get more people discovering and regularly reading those media. And in lieu of that, there is still a need to deal with the mainstream media and their often preference for catchy headlines and adherence to the five Ws of journalism (who, what, why, where and when – as opposed to the five Ws of tabloid journalism: wow, wacko, wizzer, whoopee and wham!)

And that takes some skill and training in working with the media.

Guidelines for communicating with the media

The UK's Social Issues Research Centre has put together an excellent guide for scientists on communicating with the media, with the aim of encouraging effective engagement and dialogue on science and research.[22] It states it is crucial that scientists understand the role of the media, and the importance of it obtaining advertising or audience, when they are seeking to spread news about their research. The guide offers a number of key points, summarised here.

Why even talk to journalists?

There is a common misperception that the media is the 'enemy' of the science community and is always looking for an opportunity to criticise the work of researchers and to hold them accountable for many of our societies' current ills. However, this is actually a minority opinion. The more general consensus is that the media play an important role in communicating science to the public and are crucial to the wider process of dialogue and engagement.

Read the papers, watch the TV!

Be aware of how science or your field of science is being covered in the media, and who is covering such stories. What are the main issues and areas of debate that are highlighted? Are scientists portrayed as divided? There is, after all, little justification for being surprised when journalists pose questions about an area of research that has been covered in previous reporting.

Get to know journalists and the world of journalism

Styles of journalism vary enormously in different countries and across different media and by different individual journalists. Be aware of these differences so you can use them to your advantage, not be disadvantaged by them.

Use your media office or take media courses and exchange schemes

Media offices are staffed with professionals who have been employed because they know the media in the same way that scientists have been employed because they know their science.

If you want to know more science, you go and study it. So if you want to know more about the media – likewise go and study it. There are many short, medium and longer courses you should be able to find.

You can also find media training courses run through your local Science Media Centre.

What is the status of my research?
Make it clear if your research has been published in a peer review journal. But don't expect journalists to be as impressed by peer review publication as your peers are.

What's new?
This is a big one. While there is a natural tendency for scientists to emphasise what is *novel* about their research findings, journalists are more interested in what is new about them. In a news story, a five-year research project's findings can sometimes have equal weight to an off-the-cuff comment.

News needs things to be new.

Risks in context
Because, as we will see in Chapter 18, the public perception of risk can be very different from a scientific view of risk, it is important to put risks in context. If the journalists or broadcasters you are talking to are not clear about the implications of your work, the potential for wider public misunderstanding is greatly increased. More so as the idea of risk is a good one for a media story, so you will need to ensure the journalist does not overplay it out of context.

A good way to put risks into context is to compare them to an existing risk that might be more familiar to people.

A researcher's advice on working with the media

UK researcher and blogger Emily Porter undertook a media training workshop with the UK's Biotechnology and Biological Sciences Research Council, and described her top five lessons as:

1. **Prepare.** Think before you speak. Sign up for media training at your university or institute if they offer it. If not, find a media course.
2. **Structure.** If you are going to write a media release, understand the requirements of the 5 Ws for your first paragraph (who, what, when, where, why).
3. **Language.** Use words that a wide audience can understand. Avoid jargon.
4. **Share.** Social media isn't scary. If you don't have a million followers on Twitter, don't worry: tweet or message other researchers who do have large followers and they may retweet it for you.
5. **Personality.** Be animated when doing interviews, but also in situations such as writing a blog post.[23]

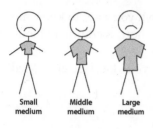

Small medium   Middle medium   Large medium

The importance of finding just the right medium cannot be over-emphasised

Writing a good media release

So what kind of book on communication doesn't have a section on how to write a media release? Of course there is going to be a section on how to write a good media release.

But first we need to consider how much the role of the media release has changed in recent years. These days releases can be tweets, Facebook posts or just text messages. The medium is not the priority. The message is.

NASA and the European Space Agency have developed a set of media release guidelines that are as good as any you'll find. They note:

- It is critically important to always tailor the style, level and content of a media release to suit the needs of the media and not the secondary target groups who may also access the release, such as politicians and interest groups.
- The media release should fulfil three main goals:
  1. to increase awareness of science and the scientific work process
  2. to increase awareness of the organisation
  3. to increase awareness of specific scientific projects, instruments or missions.
- There are three types of releases:
  1. news releases focused on one or more scientific discoveries
  2. photo releases that contain 'pretty pictures', but no big discovery
  3. video releases, in a video format designed for use on broadcast television or on websites.
- Your story should be timely, relevant, have local interest, have strong implications and settle some ongoing debate or intrigue. It should also have human interest, significant science with a new or interesting angle. And beautiful images.

Science Media Centres

There are Science Media Centres in the US, Australia, the UK and New Zealand for starters. They do an excellent job of linking scientists and the media together and are well worth discovering. They generally offer four key services:
• Rapid response to media requests for interviews with experts, background information, or other data and analysis for fast-breaking news stories.
• Advance briefings and alerts to keep journalists up to speed on breaking science and technology news and discoveries.
• Resources including fact sheets, issues briefs and reports about important science and engineering topics.
• Training for scientists, engineers and health professionals to enable them to offer reporters the facts and comments that journalists need, in the form they need them.

Useful communication and media resources

- Messenger: Media, science & society, engagement & governance in Europe
  http://www.sirc.org/messenger/messenger_materials.html
- Communicating EU research and innovation
  http://ec.europa.eu/research/participants/data/ref/fp7/146012/
  communicating-research_en.pdf
- Communicating science: A scientist's survival kit
  https://publications.europa.eu/en/publication-detail/-/publication/1117e636-c60e-4241-
  9c1e-9aaaef5a59bd/language-en
- Science Media Savvy, a free training resource that provides tips for scientists on working
  with traditional and social media
  www.sciencemediasavvy.org
- Practical advice for press officers and scientists wishing to communicate
  https://www.eiroforum.org/wp-content/uploads/200511_tips-press-officers-science-
  communicators.pdf
- NASA/ESA, Press release guidelines for scientists, available on the European homepage
  for the Hubble Space Telescope
  http://www.spacetelescope.org/about_us/heic/scientist_guidelines.html
  (Source: http://www.sirc.org/messenger/)

- Images, illustrations and visual design are key factors in successful science communication and without good visuals the chances of getting your story across diminish – particularly in an age of decreasing attention spans.

And add to that some standard advice for science writing in general:

1. **Prepare properly:** Know your story (based on the six golden questions: What? When? Where? Who? Why? [and How?]).
2. **Do your research:** Scan the current scientific literature on the topic and mine the web.
3. **Structure your thinking:** Brainstorm the topic to help you choose your angle.
4. **Simplify:** Make texts as simple as possible. Nowadays people simply do not have time for lengthy explanations.
5. **Explain:** But do explain when needed. Particularly acronyms and scientific terms.
6. **Edit:** Re-reading and editing a text always improves its quality.
7. **Use your software:** Spelling and grammar checking in word processing packages make this easier, but don't forget to look for words spelled wrong that are another actual word (e.g. bowl and bowel, prostate and prostrate and public and pubic!).[24]

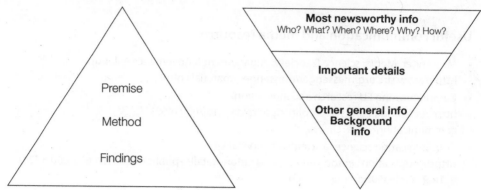

Fig. 10.5. The inverted pyramid. On the left is how scientists tend to structure a story, and on the right, how the media structures a story.

And never forget the good old inverted pyramid, shown in Fig. 10.5. Take your abstract and turn it upside down. Rather than starting with the methods and so on – start with the most important thing – what you found, and then add the next and next most important things.

A final important word: Mistaking media coverage for impact

Getting into the media is not actually the end game – getting to your audience is. It is very easy to confuse the two, by measuring the amount of media coverage you got without actually measuring how many of your articles or interviews were seen by, and impacted upon, the audience you were seeking to reach.

Getting a major story in a newspaper doesn't do you as much good as you'd like to think it does if it is only really seen by the budgie whose cage it is lining.

According to the 2018 Edelman Trust Barometer, a huge amount of people get their news through technology platforms, which we know are often self-referential echo chambers that enable beliefs to equal truth. This has made it harder for institutions like the media to play their traditional roles and has created a lot of confusion, with 63% of people surveyed saying that that the average person doesn't know how to tell good journalism from rumours or falsehoods.

But they do suggest a solution. They say that in the modern world content is no longer enough. It has to be now be tailored for the algorithms that dominate the flows of information, and to do that it has to be fit for purpose so that it can be easily shared by the 25% of people who amplify content out to the broader public. Steve Rubel of Edelman describes this as information that is personal, visual, informal.

Case study: A cautionary tale about getting the media release right

In 2017, leading biomedical researcher Professor Sally Dunwoodie and her team were ready to release the findings of a major medical discovery. Experimenting on mice they found that if pregnant mice were deprived of vitamin B3, or niacin, they were more likely to have birth defects and miscarry. Yet when it was added to their diets the birth defects dropped.

They knew they had found something significant and their research paper was accepted by the prestigious *New England Journal of Medicine*.

The institute she worked for, the Victor Chang Cardiac Research Institute in Sydney, did the standard thing and put out a media release to better promote the story and the Institute's work. They issued the release the same day the journal published the research, in August 2017. It was headed 'Historic discovery promises to prevent miscarriages and birth defects globally'.

The executive director of the Institute, Professor Bob Graham, the Institute's Board and Professor Dunwoodie all signed off on the release. It stated plainly that Vitamin B3 could 'cure molecular deficiencies which cause miscarriages and birth defects'. It also stated, 'Every year 7.9 million babies are born with a birth defect worldwide and one in four pregnant women suffer a miscarriage in Australia. In the vast majority of cases the cause remained a mystery. Until now.'

The world's media jumped on the story and Sally Dunwoodie not only appeared live on the BBC but was getting calls from the US, India, China, the UK and Germany. It was a great story.

So far so good.

But then things turned around. Other medical researchers started being quoted in the media questioning the findings and branding the announcement as 'harmful'.

What happened?

It wasn't the study that was actually being questioned, so much as the release. For instance, it had originally stated in part, 'using a preclinical model' – which was later changed to 'using a preclinical mouse model.' Indeed, nowhere in the original release was a mouse even mentioned.

It was later revealed this was done because of a fear of upsetting the animal liberation lobby. Instead they upset the medical and scientific community.

Many felt that the claims were overstated because no human clinical trials had yet been undertaken.

It was a harsh lesson for the Institute that hyping a story can be a two-edged sword – it can get you more coverage, but can lead to accusations of being misleading.

The final verdict, summarised by the editor of *The Medical Journal of Australia* was that the science was impeccable but the way it was sold was overblown.[25]

Useful and not useful comments

Sometimes a journalist or science journalist might contact you and ask for your opinion on a study that someone else has conducted. Science journalist Ed Young has given a short list of things that he finds useful and not so useful when trying to get comment.

Useful:
- Comment on the strengths and weaknesses of a study and whether it found something really new.
- Give your reaction to the study.
- Tell how it fits with previous studies and if anyone else has found similar things.
- Use simple language.

Not useful:
- Summarising what the paper showed, as is provided in the abstract.
- Using vague adjectives like 'interesting'.
- Saying, 'This research is interesting but more work needs to be done.' – it's the most banal quote ever.
- Playing publication politics or citation politics on why it was published where and who was cited and who was not.[26]

What to do with what you now know

The next time you have to write a media release, or do a media interview, do your research. Know what messages are likely to work best for particular media and their audiences.

If you are not sure, seek out professional advice. There are lots of people who have spent many years studying and working in media relations positions, or whose role in life is to help make you and your message more media-friendly.

And try and be timely. As Fiona Fox, Director of the British Science Media Centre put it, 'The search for truth and respect for evidence and accuracy that drives the pursuit of knowledge by scientists is about as far removed as it could be from the media's needs at times of breaking stories ...' (Even though, she adds, it is exactly because of this integrity and respect for evidence that she wants the public to hear from that expert.)[27]

Sometimes you do need to dive in a little unprepared – but arm yourself with that great answer that is used in job interviews, 'No, I don't know the exact answer to your question, but I know with confidence where I would find the answer.'

And also, treat every encounter with the media as a learning exercise to build on.

Key summary points

- The media landscape *is* complex and can change rapidly.

- There is a complex relationship between the mainstream media and new media, with the impact of both often overstated.

- Knowing how the media works increases your chances of getting your story run in the media.

- Don't make the mistake of confusing media coverage for impact.

11

Being a social media superstar

'Social media is clearly driving us all insane. This is a psychological experiment that nobody signed up for and we are all in it, and we each have to curate the contents of our own consciousness a little more carefully than we have been.'

– Sam Harris, neuroscientist

It is a risky proposition writing anything about social media, as it goes out of date so quickly. But we can look at what is known and make some brave predictions. There is a change happening around social media – or new media – with polarising levels of trust, not just in messages, but in mediums. Events like the Cambridge Analytica scandal (whereby a private company that was working for the Trump presidential campaign was found to have been given access to the data of about 87 million Facebook users) have diminished trust in Facebook. But only among some.

Battles for legitimacy between online agencies have led to people seeing an equivalence between media such as *The New York Times* and a website like Breitbart.[1] But only for some.

There is growing concern that social media allows for the flourishing of anti-science ideas, with online algorithms that keep feeding you the same misinformation or dodgy science on autism, cancer cures, climate change, alien invasions and so on – the same way it allows you to only see things that support your political ideology. But only for some.

There is concern that social media is so non-discriminatory and non-fact checked, that a 22-year-old conspiracy theorist still living at home with his or her mother can be accorded the same standing as a professor with 30 years of experience. But only for some.

There is also a worry that diminishing any complex message to the 280 required characters of Twitter has led to a general dumbing down – seen in the politics of policy by tweets (you know who we are talking about!), which also diminishes the complexity of science messages. But only for some.

There are scientists and science communicators who still see social media as difficult and dangerous terrain and only dabble with it around the edges, presuming that will be sufficient. But only some.

And there are people who will read this chapter and think, yeah, but I know all that already and more, and I don't need such basic level information. But only some.

#giveitareadanywayandsee

Is social media changing our brains?

Professor and Baroness, Susan Greenfield, neuroscientist and member of the House of Lords, believes that digital technology is actually changing the human brain. She has said that the need to respond more to fast-paced information is changing the default consciousness settings for younger people, adding to the existing digital divides.[2]

Already there are clearly:

- digital natives – who are born using and understanding digital technologies
- digital as a second language – who acquire understandings (but let's be frank, talk it with an accent)
- digital immigrants – who struggle with it, or keep up – just
- digitally dysfunctional - who just don't get it and may never do.

The general trend is that the younger a person is, the higher up that list they will be and the older they are, the more likely at the bottom of the list they will be – which happens to mirror the general management structure of organisations, making it hard to have a digital engagement strategy that works at all levels.

This can be as difficult a challenge as it was trying to introduce computers to organisations that were wedded to the dictation and typing pools – that ended up with many senior managers one-finger typing memos and wondering how this new-fangled technology was ever supposed to be more efficient.

So who is using social media?

Where is a scientist or science communicator to dip their toe into the pool of social media? It may help to know that more scientists use Twitter than Facebook to share their science.[3] But if you want to reach the public, you need to be aware of what platforms are being used by whom.

According to a 2018 study by the Pew Research Center, the most popular form of social media in the US is YouTube (used by 73% of people) followed by Facebook (68%). But when you look at new media use by age, you tend to see younger people have stayed with YouTube but are abandoning Facebook (see Figs 11.1 and 11.2). Some have put this down to a feeling that the older generations have invaded Facebook and made it boring, but there are also instances of people who have been embarrassed by the 'What goes on Facebook *stays* on Facebook' issue, while other mediums like Snapchat don't store your photos.

Young people, especially those aged between 18 and 24 prefer to use Snapchat (78%) and many people visit it many times a day. Slightly fewer, 71%, use Instagram and almost half (45%) are Twitter users. But for young people, the biggest platform is YouTube (94%).[4]

So what is the main difference between Twitter and Facebook?

Facebook is for following everyone you went to school with, and Twitter is for following everyone you wish you went to school with.

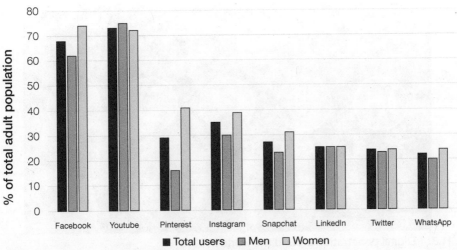

Fig. 11.1. US social media use demographics by gender[4]

They also found some interesting gender and ethnic divides in use:

- Pinterest is more popular with women (41%) than with men (16%).
- LinkedIn is more popular among college graduates and those in high-income households (about 50% of Americans with a college degree use LinkedIn, compared with just 9% of those with only a high school diploma or less).
- WhatsApp is popular in Latin America and South East Asia (which tends to extend into those migrant communities too).[4]

Australian studies have shown:

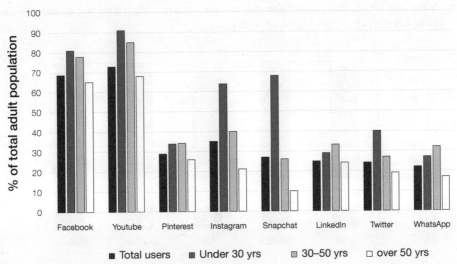

Fig. 11.2. US social media use demographics by age[4]

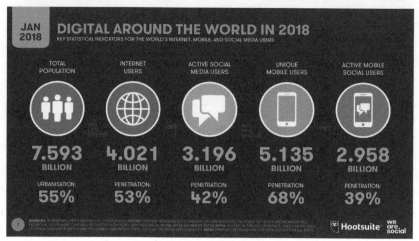

Fig. 11.3. Digital penetration around the world in 2018[6]

- About 60% of the population are active users on Facebook, with about half the country logging onto Facebook at least once a day. By age, the largest number of users are in the 25–39 years bracket.
- About 60% also watch a YouTube video at least once a month.
- The third most popular platform is Instagram, with about 37% of people using it, followed by WhatsApp (20%). LinkedIn and Snapchat are both used by just under 20% each.[5]

As for scientists themselves, statistics differ enormously, but include figures like:

- 77% of life scientists participated in some type of social media
- 50% saw blogs, discussion groups, online communities, and social networking as beneficial to sharing ideas with colleagues[8]

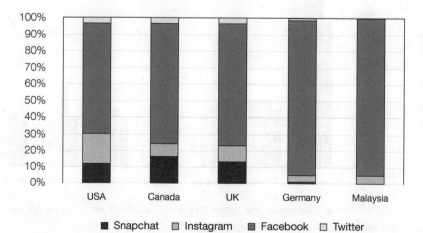

Fig. 11.4. Time spent on different social media in different countries[7]

- 41% regularly use LinkedIn, 38% regularly use Facebook and 13% regularly use Twitter – though as was previously stated Twitter is more popular to share science information.[9]

The problem is always THEM, not US

Perhaps the most worrying trend for scientists and science communicators is not how well or not digital technologies are being used within an organisation, but how they are being used outside of it.

Sir Peter Gluckman, the very articulate former chief scientist of New Zealand, has said that open access to knowledge has led many to believe that there is no need for expert interpretation of information. And yet as Google has become the almost universal source of 'facts' there is no way of knowing what is reliable or unreliable without expert interpretation.[10]

My wife passes all kinds of crazy cancer cures and cancer cause stories to me that her wider Chinese-descent family pass around on WhatsApp. One day she passed one to me about cancer treatment derived from a plant found in tropical Queensland. I said, 'Yeah, yeah. Sure I'll look at it.' (You know that tone, right?) When I did look at it, I found it was posted by a university, not some dodgy private clinic operating out of Wherethehellistan. And that had me thinking, sure, I can tell the difference in content – because, well, because I insist that the medium I am receiving things from conforms with my biases before I even look at the message. But how many people are swayed by the message over the medium?

Steve Sloman and Philip Fernbach have pointed out in their book *The Knowledge Illusion*, the importance of peer groups in creating collective knowledge.[11] But if peer groups are largely formed from our social networks, and are self-selected for their similarities of view, it drives debates to polarised extremes rather than any sense of compromise and consensus.[10]

And as science communication guru Dietram Scheufele states:

> Given that only 16% of the US public reports in surveys that they follow news about science 'very closely' (down from 22% in 1998), an emerging media landscape that further tailors searches and news toward these popular preferences rather than the types of (scientific) content that citizens need to make sound policy choices is at least somewhat disconcerting.[12]

But let's be frank. This is just the environment we are living and operating in, and we need to look for ways to work with it, rather than decry it if we are unable to easily change it.

The benefits of using new media

Social media tools offer a powerful way for scientists to boost their professional profile as well as act as a public voice for science. Scientists are increasingly using social media to share journal articles, to advertise their thoughts and scientific opinions, to post updates from conferences and meetings, and to circulate information about professional opportunities and upcoming events. Therefore, not having an online presence can severely limit a researcher's visibility.[13]

They recommend several key strategies that you might benefit from, that include:

- **Establish a professional website**
  At a minimum all researchers and science communicators should set up a personal website that lays out their specific research projects and areas of expertise, that can be found by simple online searches.
  Supplementing your website with social media accounts (e.g. Twitter and LinkedIn profiles), will help you appear higher in online search results.
- **Locate relevant online conversations**
  Find people with common interests and follow the social media that they follow. You can read online conversations without contributing to them to learn the basic etiquettes of different social media platforms.
- **Navigate the deluge of online information**
  Try not to drown in it all. Explore multiple social media tools and related sites or apps for managing online accounts. Though popular content is often heavily reposted and shared, the most important articles and conversations will usually reach you at some point. And don't be afraid to ask for help – there are many friendly and established communities who are willing and eager to assist new users.
- **Interact with a diversity of participants**
  To be effective you need to engage with audiences online, but don't just engage with people who think like you. Be open to new discussions outside your own professional area.
- **Reach your audience**
  Online science communication channels are probably only reaching people who are interested in talking about science online. Don't expect to reach people in the way mass media does.
- **Use online tools to improve research efficiency**
  Online communities can be especially useful for niche topics where community members may have specialised skills or knowledge that can help you.
- **Use online visibility to help track and improve scientific metrics**
  There is mounting evidence to suggest that an active online presence may directly impact a researcher's credentials, as they are measured through traditional metrics. One UK researcher, Melissa Terras, observed that tweeting and blogging about her own papers led to spikes in the number of article downloads.[14] And highly-tweeted journal articles were 11 times more likely to be highly-cited versus articles without strong social media coverage.[15]
- **Enhance professional networking**
  Conversations on Twitter or Facebook can serve as an informal introduction to meet other scientists at a conference.
- **Improve communication with the public**

Understanding online tools

To help scientists understand how best to use online platforms, US researchers Holly Bik and Miriam Goldstein have provided a quick reference guide to the most common tools and how they are used.[13]

Blogs. Short for web log. Traditional, long-form online narratives. WordPress (http://wordpress.com) and Blogger (http://blogger.com) are two of the most popular sites to offer free blog hosting. If you don't have the capacity to blog regularly though, it can be better to ask about writing a guest post on an established blog that has an existing audience.

RSS feeds. A type of URL – or web page address – that allows users to automatically mine blog or website updates without doing an online search. RSS (rich site summary) aggregators such as Bloglines or Feedly are a simple way to keep track of new and relevant content.

Twitter. A very popular social networking site that limits posts to 280 characters. Twitter is useful for in-the-moment conversations, customised news streams, and building and maintaining communities. Hashtags, such as #asm2018, for example, allow users to follow all the tweets related to a particular event.

Facebook. The most widely used social media site. Some use this professionally for science-related content, and some prefer to only use it for personal content.

Tumblr. A microblogging site that can publish any type of media very easily and quickly. Users post photos, videos, or short quotes as opposed to long written narratives.

Pinterest. A photo-only microblogging site where users can define their theme (e.g. food or science).

Storify. An app that can organise your tweets, videos, blog posts and other media. For example, if there is a panel discussion or academic seminar, a Storify can be created that includes live tweets from the audience, videos of the panellists, and links to their publications, websites and social media profiles.

Linking communities. Digg, StumbleUpon, MetaFilter and others are content aggregation sites that recommend new and interesting content to subscribers.

As an example, two projects aimed at changing the perception of science and scientists themselves have recently gone viral in the online science world: the hashtag #iamscience and *This is What a Scientist Looks Like*. These initiatives were designed to raise scientists' profiles, dispel stereotypes, and highlight the unconventional career paths followed by most scientists.[13]

If you are going to do it, do it well

Paige Brown Jarreau, from the Louisiana State University's College of Science, recommends several key tactics for social media success. They include:

Eggs = Milk Cheese^2

Albert

Huh?

Annalen der Physik, Ed

Albert Einstein struggles with text autocorrect

Using social media

The Australian Science Media Centre provides some useful advice on how to best use social media, which includes the very important tip of not paying too much attention to negative comments. That is a part of the online world that allows for and encourages dickheaded behaviour. Learn to accept it the way you accept that walking through the countryside you will always have to dodge some bullshit.

They advise, when communicating through social media:

✗ Don't just talk at people – aim to actively engage with them.
✓ Ask questions to encourage interaction and discussion.
✓ Interact with other pages and people (comment, share and retweet).
✓ Respond politely and respectfully to comments. Sometimes it is best to just ignore.
✓ Maintain your professionalism. Don't let your emotions rule when posting or responding to comments.
✓ Use a spell checker – it only takes a mintue – sorry, a *minute*!
✓ Be consistent: check your site regularly and build a cohesive social media presence.
✗ Don't post sensitive or confidential information – if in doubt leave it out.[16]

- Follow the people you are trying to reach.
- Have a clear idea of who your target audiences are in the first place.
- Search for hashtags your target audiences might be using. Identify and follow particular social media users based on keywords and content.
- Create Twitter lists for your target audience(s).
- Create original, valuable, engaging content for the people you are trying to reach.
- Engage with the people you are trying to reach.[17]

Finding the right medium

Over the past few years, the US National Academies of Sciences, Engineering and Medicine have held a rather impressive colloquium (yeah, I know, it's a big word for a meeting) on the Science of Science Communication. At the 2017 meeting, Gerald Davis from the University of Michigan told a story about his attempts to experiment with several ways of communicating the findings of his research on the threat posed by vanishing corporations.

- Cranial Facial Book
- SecondPersonSingularTube
- Pipetterist
- $4^2 = 16$
- Bird vocalisation
- Snapdialogue
- TheMissingLinkedIn
- Instamilligram

Scientific social media platforms that just never took off

He said that he first published an academic book with a university press – but such books serve mostly to convince your colleagues.

Then he wrote a mass market book – but that only works 'if you want to convince your parents,' he said.

He appeared in segments on the Public Broadcasting System's NewsHour and National Public Radio – but the audiences for those programs are limited.

He then wrote short pieces for an internet outlet – but this was unsatisfactory, especially when he began to read the many ill-informed and disparaging comments generated by his articles.

He toyed with Twitter – but said that while powerful incentives still exist to focus on research that yields short and pithy findings, he was worried that moving to do more Tweet-sized work in bite-sized chunks was not the same as doing real science.

He was enthusiastic however, about the impact of short and accessible pieces written for reputable and intelligent outlets like *The Conversation*,[18] whose articles, it should be noted, are often retweeted, emailed, posted on Facebook and cross over into the mainstream media – so it can be considered something of a social media multiplier. So there can be a big advantage from finding social media platforms that act as multipliers and do a lot of the work for you and save you from posting endlessly to a ghetto blog that has fewer than 10 readers – including your parents.

For information to be best able to be disseminated by multipliers it should be presented in a style that is consistent with digital culture, being personal, visual, informal and, above all, emotional.[19]

Lessons from NASA

NASA has one of the greatest reaches of any science agency with its social media, and yeah, it's pretty easy to get a zillion followers when you have pictures of Mars and astronauts and things. But they actually have a pretty good social media strategy to support their quality of content. So ideally you need both – good content and a good strategy.

Four communicators – Amy Mollett, Cheryl Brumley, Chris Gilson and Sierra Williams – have outlined the things they have learned at NASA into a book, *Communicating Your Research with Social Media*.[20]

They said that where television documentaries and public exhibitions were once relied upon, social media platforms have now brought new opportunities for scientists and communicators to interact with their audiences.

In 2009, NASA astronauts sent the first tweet from space. In 2010, they did the first Foursquare check-in from space. In 2011, they were the first US Government agency on Google+; and in 2012, they had the first Foursquare check-in from Mars.

The authors said that while they have witnessed a great deal of optimism and legitimate backlash about researchers engaging online, much less attention has been directed towards looking at where social media fits into a researcher's communication needs, and how it might be employed more constructively. And rather than be overly prescriptive about what social media you should use and when, they wisely advise:

> Given the particular nuances and requirements of the modern-day research environment ... social media is meeting very particular research needs in very particular ways. But being aware of these specific needs for your own research project will help you decide where to invest your energies and what types of digital content may be worth experimenting with.[21]

Can you have too many tweets?

Ecology writer and science communicator Ian Lunt talks about 'peak tweet', the same way we talk about peak oil, or peak TV (or as he says, 'peak beard' – which is the point at which the rate of beard production is exceeded by the rate of beard destruction). He says that peak tweet is the time when the rate of tweet production far exceeds the rate of potential consumption.[22]

He also says that he experienced peak tweet way back in the distant past of 2015, at a sizable ecology conference, where attendees live-tweeted far more messages than readers could find or read.

Case study: Should social media sites be responsible for the misinformation they allow?

In a comment article in the awesome online journal *Undark*, science writer Michael Schulson asked if Google and Facebook should be responsible for the 'medical quackery' that they host. He has argued that social media algorithms help charlatans spread autism cures, vaccine disinformation, and AIDS denialism through online videos and other mediums.

He tells the story of Dr Biswaroop Roy Chowdhury, an Indian engineer who claims he has an honorary PhD from Alliance International University in Zambia (which the Zambian Higher Education Authority deregistered in February 2018 after it failed an audit).[25.] Anyway, Chowdhury posted a video on YouTube in early 2018 arguing that HIV is not real, and that anti-retroviral medications actually cause AIDS. The video received over 380 000 views within a few week, with more watching it on Facebook.[26]

Chowdhury's story is not unique, as many charlatans, hate mongers, conspiracy theorists and junk science peddlers have successfully used online platforms to find supporters for their dubious causes.

'Potentially deadly cures for autism have found sanctuary on social media for years,' Schulson says. 'Desperate cancer patients have been lured online to baseless treatments peddled by shady 'experts'.'

So he asks, when social media platforms allow people to be nudged towards dangerous medical decisions, should they bear any responsibility?

The notion that AIDS is an elaborate hoax has persisted in some corners of the globe since the disease first emerged widely in the 1980s. In South Africa, the country's former President, Thabo Mbeki, was an AIDS denialist, and his policies contributed to an estimated 365 000 deaths, according to Harvard researchers.[27]

Seth Kalichman, a psychologist at the University of Connecticut and an expert on AIDS denialism, has said of denialists, 'Their visibility is increasing even as they become less and less relevant. I think they are fizzling out and if not for social media, they'd probably have become a thing of the past.'

More worrying, Schulson says, is that the digital platforms that host this type of dodgy material have algorithms that give visitors more of the kind of content that they look for, perpetuating beliefs in fringe medicine.

And how do platforms respond to such concerns? When Facebook was challenged over some of the content they ran or supported their response was to admit that contentious perspectives exist on their platform but that 'we have long believed that simply removing provocative thinking does little to build awareness around facts and approaches to health'.[26]

Some of the best science YouTubes to get hooked on

The most popular science channel on YouTube at the time of writing was Vsauce, with over 14 million subscribers in early 2019, followed by AsapScience with 8.5 million, and then a bunch of others jostling for third place with around five or six million subscribers. But rankings and subscriber numbers can change. More helpful is to follow the tone and content of each channel.

Vsauce. Now spun-off into Vsauce, Vsauce2, Vsauce3, WeSauce and Dong (Do Online Now Guys – not what you might have thought!) The channels concentrate on videos on scientific, psychological, mathematical and philosophical topics.

AsapScience. Uses drawn animations and covers great common interest topics.

SmarterEveryDay. Destin Sandlin talks like your friendly neighbour next door and explains how things happen or work all around us.

Kurzgesagt (In a nutshell). Uses animations to explain complex science.

SciShow. Posting almost daily, this is where you can learn about slime moulds, and the science of beer.

Veritasium. Derek Muller walks you through many science concepts in a very plain-speak style.

Minute physics. Often longer than a minute, but a great channel for information on physics.

It's okay to be smart. Want to learn about the science of Game of Thrones? This is the channel for you!

Periodic videos. Concentrates on chemistry.

Home science. Science stuff you could do at home.

BrainCraft. Everything you might want to know about our brains and how we think.

Numberphile. Principally about mathematics.

IFLScience. Spun off from the very popular Facebook site. (Check it out to find out what the acronym stands for!)

Generally, however, tweeting a conference is seen as a useful thing because people tweet for a variety of reasons, including:

- they are typing up notes anyway, and it is just as easy to tweet them
- they want to reach other tweeters at the conference and contribute to the conference 'buzz'
- they want to reach a wider audience of researchers or the public.

Some tips Ian gives for using social media are:

- Measure your success in shares, not reads.
- As more and more readers view the web on a mobile phone, you need short paragraphs, and lots of images. Every page scroll kills a reader.[23]
- You don't just need to learn how to sum up complex science in 500-word grabs, or two-minute videos and podcasts; you need to do it so the audience don't realise what you have done.
- Every post must stand alone, to be read in isolation or in a random order.

- Start with something attention grabbing. Give the punchline first.
- Throw out all your excessively grandiose big words.[24]

Thinking outside the digital box

The most popular YouTube channel in the world in 2018, according to Wikipedia, was a comedy channel called PewDiePie. The channel has more than 60 million followers and features seemingly random chats to camera that grew out of game walk-throughs. PewDiePie (Felix Kjellberg) has been described by *The Guardian* as 'Childish, offensive and immensely popular' – but also as 'funny, intelligent, innovative and highly-charismatic'.[28] The challenge might not be to find ways to get other science videos as popular as PewDiePie – but how to get him to talk more about good science. After all, he has already started a book club.

In that vein, the New Zealand Prime Minister's Science Communication Prize in 2017 was very rightly awarded to Damian Christie, who had the brilliant idea to take a young lifestyle Vlogger, Jamie Curry, down to Antarctica. She posted YouTube videos to her large base of followers – predominantly young women aged between 13 to 25 – about issues like climate change and climate science. The series of videos, known as *Jamie's World on Ice*, received over 2.5 million views – among an audience who would not normally be science followers.[29]

A word about gender-based trolling online

It is definitely worth mentioning the big, fat, ugly, man-child gorilla in the room – that for women engaging in social media they are often a magnet for: big, fat, ugly, man-child gorillas, who feel they need to post derogatory comments about their looks or make sexist comments about them.

A couple of studies have even quantified this. One, undertaken by Inoka Amarasekara of the Australian National University, looked at over 23 000 comments left on videos about science and found women get more critical comments than men, and more remarks about their appearances. In fact, 14% of comments on channels hosted by an on-camera female were critical, as opposed to only 6% for those hosted by a male. Also, about 3% of comments were sexist or sexual, compared to only 0.25% for men.[30]

Similar findings were made by an analsysis of TED talks in 2014.[31]

Vanessa Hill, who hosts the popular YouTube channel BrainCraft, with about half a million subscribers, described online sexist trolling as being like 'someone leaving a Post-it note on your desk every day telling why you're not qualified or why your voice is horrid.'[32]

How to make a great YouTube video

Hundreds of hours of video are uploaded to YouTube every minute – and some of it has science content. Not much, as the most popular YouTube videos are still largely of rock stars, game walk-throughs and make-up tutorials. But of course, viewers self-select, and you've about as much chance of getting a 15-year-old gamer to watch a science video as you have of getting a scientist to watch a make-up tutorial.

Regardless of how good your video is, you will only ever have about three seconds to connect with a viewer before they choose to stay watching or jump off to something else.

Australian researchers Dustin Welbourne and Will Grant analysed nearly 400 science communication videos for research (though don't let your kids get away with telling you all their YouTube watching is for research!) and they found seven things that define a successful science YouTube video, that are likely to be successful if followed.

1. Choose your audience
All different videos are aimed at a different audience. Know your target audience and stick with it. Channels with lots of subscribers generally stick to one version of science communication.

2. Deliver in style
Delivering information on video can take many styles. Generally, these formats fall into:
- a vlog (video blog)
- a voice-over animation
- a recorded presentation
- an interview.

3. Stay focused
Stick to your theme and topic.

4. Get to the point
If your video fails to hold the viewer's attention, there are plenty of other videos that will, so get to the point quickly. And talk faster. Talking to a live audience at up to 150 words per minute is fine, but for a YouTube video, aim for about 180 words per minute. And try and keep your video under five minutes.

5. Be part of the community
YouTube is a participatory culture, and that means you need to be part of the community if you want to grow your channel. This means interacting with both the consumers and creators of content.

6. Give the audience an anchor
The most popular videos have a consistent presenter that the audience can relate to.

7. Be a person not a company
Those YouTube channels that operate as extensions of existing brands (for instance, universities or TV broadcasters) are significantly less popular than channels that feature a

single person presenter that are born and grown on YouTube. So bring some personality to it.[33]

Of course, if you are thinking of starting up a YouTube channel or becoming a YouTube superstar, it's worth noting that it is quite a crowded marketplace, with well over 100 science YouTube channels out there already. You need to ask how yours will be different (and consider there already is one talking science from a bathtub!). Then you need to set realistic growth forecasts and be prepared for a long haul to get followers.

And of course, there are even lots of YouTube videos out there on how to start a YouTube channel.

What to do with what you now know

If you are serious about engaging in social media, do it well (and it is increasingly hard to not be serious about social media). There are, unfortunately, too many people half-heartedly engaging in social media who then criticise it for not getting much flow-on for them. If you don't use it often and don't know the language and norms used on different platforms, it will show.

If you are not great at it, learn from those who are already doing it well, or get someone who is born digital to advise you. It is a complex and rapidly changing field that can suck up a lot of your time for little response if not done right.

Seriously, there is nothing that is quite so embarrassing as earnest attempts to use social media by people who should have stamped on their file 'Not be let near social media!'.

Key summary points

- Social media is strongly established as a key source of information for many people and might better be thought of as 'new media'.

- Social media has changed the way information flows work, and while providing instant access to lots of great information, there are concerns that new media allows for increased reinforcement of fringe ideas with no reference to experts.

- If engaging in social media, understand the requirements, usage and followers of the different platforms.

- Regardless of the platform you choose, you will need both good content and good strategy.

- And if you are going to engage in new media – do it well!

12

I've been framed! The art of framing

Michael Scott: 'I've never framed a man before, have you?'
Dwight Schrute: 'No. I've framed animals before.'

– The Office, 2005

Framing is a useful thing to understand well. And if done really well it can be a very powerful communication tool. But if not done really well it can be a bit – well – embarrassing.

Let me give you an example. You come home one day with a new car that you haven't told your partner you were going to buy, as it was a bit of an impulse purchase. Do you say, 'This is going to be so good for our family as it is a much safer car.' Or do you say, 'But look how red it is, and look how fast it goes, and the salesperson was really nice to me.'?

Put simply, framing is the way you tell your story. The angle, if you like. The type of analogy and anecdote and values that you use. The car is still the same car – but there are different ways of considering it.

The US National Academies of Sciences report *Communicating Science Effectively: A Research Agenda*, says, 'Framing is casting information in a certain light to influence what people think, believe, or do.'

The contentious word there is 'influence' as this can be interpreted in different ways. I've heard some people argue that this interpretation of framing could be seen as slanting the facts, or manipulating data, and the only really, purely, unbiased way to reach an audience is to just tell them the facts.

But if you've been paying close attention to the chapters before this one, you should realise by now that research shows that factual information is no better at influencing people than information with no factual basis whatsoever.[1] Yes, cherished and hard-fought-for scientific facts don't tip the scales very far when they are weighed up with rubbish facts.

What tips the scales is how the facts are presented, and if that aligns with a person's particular way of thinking.

That's framing.

If you know your partner thinks your current car is perfectly good, it is going to be very hard to convince him or her that your new red car is really necessary. But if your partner believes a new car will make your lives more meaningful and enjoyable and you'll drive around the coast with the windows down laughing like you are in an advertisement for new cars, then it will be fairly easy for him or her to see it as a good purchase.

Or as George Lakkoff puts it: 'People think in frames … to be accepted, the truth must fit people's frames. If the facts do not fit a frame, the frame stays and the facts bounce off.'[2] The significance of this idea cannot be understated, as it indicates that engagement activities that are based on informing and educating an audience with strong existing views may have very little impact.[3]

If you take climate change, for example, there are many different ways it can be presented. It can be seen as a severe environmental risk, or as a public health risk, a challenge for farmers and growers, an increased fire risk, or it can even be seen in a more positive light as an opportunity for innovation and economic development.[4]

In fact, three researchers, led by Sander van der Linden (who was named a 'Rising Star' by the Association for Psychological Science in 2017), looked at causes for disengagement on climate change issues – such as not perceiving it as urgent nor relevant to them – and came up with five key frames for climate change that improve engagement on it. They were:

1. Emphasise climate change as a present, local, and personal risk.
2. Facilitate more considered and experiential engagement.
3. Leverage relevant social group norms.
4. Frame policy solutions in terms of what can be gained from immediate action.
5. Appeal to intrinsically valued long-term environmental goals and outcomes.[5]

But of course the challenge can be in getting the right frame to align with the right audience. If you look at Genetically Modified Organisms (GMOs), information that is framed in terms of social progress and improving the quality of life may well fit closely with one person's way of thinking, but somebody who was more interested or concerned with public accountability would need another frame.[5]

Framing exists even when we don't know we are using framing. George Lakoff, in a paper entitled 'How we frame the environment', cites a memo to the Bush administration written by Frank Luntz, on how to win the global warming debate. The memo stated:

> It's time for us to start talking about 'climate change' instead of global warming … 'Climate change' is less frightening than 'global warming'.

Lakoff stated that the memo was the beginning of the use of 'climate change' as a popular term but based on the idea that 'change' left out any human cause for the change. Climate just changed. No one to blame.[6]

Luntz, who was also a Republican pollster, used the findings of focus groups and surveys to come up with other messages that did not align with the dominant scientific view on climate change, such as the 'scientific debate remains open', or that 'further research is needed before government action is taken', or that 'any USA policy action would lead to 'unfair' economic burden on Americans since countries like China and India were not also taking action'.[7]

You may have seen these messages repeated by anti-climate change activists or in the media. They were based on an understanding of which frames most aligned with uncertainty in people's minds. The repeating of these types of messages by conservative lobbyists and members of Congress was credited with the continued failure of the US to adopt the Kyoto Protocol Treaty.[8] Having an alternative to the mainstream story also allowed the media to apply the standard conflict structured stories, implying there was equality in the voices for and against human-induced climate change.[9]

A short history of framing

Among the earliest work done on framing that of anthropologist Erving Goffman, in the 1960s and 1970s. He described framing as 'schemata of interpretation' which allow individuals to 'locate, perceive, identify, and label' issues, events, and topics. Yeah, not quite as catchy as the term framing, I know. But he was trying for accuracy.

Goffman said that words can be like triggers that help individuals negotiate meaning through the lens of their existing cultural beliefs and world views.[7] His work was followed in the 1970s and 1980s by the cognitive psychologists and Nobel Prize winners, Daniel Kahneman and Amos Tversky (authors of the hugely best-selling book, *Thinking, Fast and Slow*). They undertook experiments to understand the risk judgements that people made and found that how a message was 'framed' had a huge impact on whether something was accepted or rejected.[7]

For instance, one of their experiments involved asking people which of two medical treatments they would prefer for lung cancer. The same statistics were presented to people in terms of either mortality rates or survival rates.

The information was given like this:

Survival frame:
Surgery: Of 100 people having surgery, 90 live through the post-operative period, 68 are alive at the end of the first year and 34 are alive at the end of five years.
Radiation therapy: Of 100 people having radiation therapy, all live through the treatment, 77 are alive at the end of one year and 22 are alive at the end of five years.

Mortality frame:
Surgery: Of 100 people having surgery, 10 die during surgery or the post-operative period, 32 die by the end of the first year and 66 die by the end of five years.
Radiation therapy: Of 100 people having radiation therapy, none die during treatment, 23 die by the end of one year and 78 die by the end of five years.[10]

The researchers found that the survival frames were strongly favoured over the mortality frames, and those who favoured radiation therapy rose from 18% under the survival frame to 44% under the mortality frame, which they summarised as:

> … illustrated by the preceding examples, variations in the framing of decision problems produce systematic violations of invariance and dominance that cannot be defended on normative grounds.[10]

Got that? If not, in plain non-Nobel Prize winning English, they found that people don't make the type of logical choices that many people assume others make, and that how a message was framed strongly influenced how people interpreted it.

Since then there has been more work on framing coming from the fields of politics, communication, linguistics, and cognitive science, among others. This has included studies into how the media frame things, how politicians and influencer groups do it and how advertisers do it.

Of interest have been some frames that were once very effective, but then fell out of favour, such as the US Atoms for Peace program of the 1950s, promoting the benefits of atomic power, with such things as using atom bombs to dig harbours![7] We might look back on such frames now and shake our head and wonder, what were they thinking? But rest assured, in a few decades from now people will be looking at our dominant frames and undoubtedly be thinking the same about us.

Matthew Nisbet has said that there is no such thing as unframed information, and most successful communicators are adept at framing, whether using frames intentionally or intuitively. This is necessary because audiences rely on frames to make sense of any issue. Journalists use frames to craft interest in news reports; policymakers use frames to define policy options and reach decisions; and all manner of experts employ frames to simplify technical details and make them more persuasive.[11]

Emphasis framing

The concept behind emphasis framing is that every issue has a complexity of different ways of looking at it. Your red car can be seen as an impulse purchase, as a more reliable form of transport, as a status symbol, as a compensation for getting older, as a more fuel-efficient vehicle, as a drain on global resources, etc. Different people will stack those statements in different priority, according to what they think is the most important.

We also know that people can only focus on a small amount of information at a time, and if there is a lot of complex information to consider, you can be pretty sure it is not all going to be considered. The question for a science communicator is, what influence do you want to have on which few facts or messages people focus on? That is the 'influence' of framing that is more important to focus on.

Emphasis framing is emphasising one dimension of a complex issue over another. It can be as simple as deciding what should be the lead paragraph for a story about some scientific research – which is what a journalist does when they retell a story. They choose what they think is the most important thing to tell first, which frames what comes next.

Emphasis framing can also be done in terms of telling a personalised story over providing statistics. For instance, looking at infant vaccination, a personalised story could tell what happened to a family who chose not to vaccinate their children when their child then got sick with a disease that other children had been vaccinated against. Or the framing could provide statistics about the risk of adverse health effects when vaccination rates drop below certain levels.[12]

Framing contentious issues

Of course, when it comes to framing contentious issues, different interest groups will all be seeking to frame information to their best advantage. Scientists and researchers will be talking about the economic and social benefits of nanotechnology, for example, while environmental NGOs might be seeking to frame the technology in term of unknown risks. This can make for a bit of a battleground of favourable frames – which worries some – but let's be honest, when has it ever been otherwise?

Emotive-based frames stick with us more.

There are some very effective catch-words that summarise a frame, such as 'frankenfoods' that sums up the fears of inadvertent consequences of fiddling with nature, or 'big oil' and 'big pharma' to capture the excesses and industry dominance of a particular field.

An important thing to know about this type of framing is that once such a frame has influenced people's views, it can be very difficult to change those views. One study even found that people who had never heard of carbon capture and storage could be influenced by uninformative arguments either for or against the technology, and – get this – those feelings persisted even after they had read carefully balanced information about it.[13]

What does this tell you? Frame early and frame often!

Negative frames can backfire by activating in people's mind the exact idea they are trying to counter. For instance, when President Obama said publicly that he had no intention of a 'government takeover', he ended up activating the government-takeover frame in the minds of those who fear that.[6]

George Lakoff, author of the book on framing, *Don't Think of an Elephant: Know Your Values and Frame the Debate*, states, 'One cannot avoid framing. The only question is, whose frames are being activated—and hence strengthened—in the brains of the public.'[6]

He also points out that if you want to counter a frame that is being given, DO NOT REPEAT THAT FRAME – even in trying to negate it, as it only gives the frame more life. He gives an example of former US President Richard Nixon, when defending himself during the Watergate scandal. He went on national TV and said, 'I am not a crook.' As a result, Lakoff says, everybody then thought about him as a crook.[6]

Some examples of common frames used in science debates include:

- Climate change. While those who oppose climate change use frames of 'uncertainty', those who support climate change counter with a frame of global warming being a 'Pandora's box of catastrophe'.
- Embryonic stem cells. Patient advocates have delivered a focused message to the public using 'social progress' and 'economic competitiveness' frames, while opponents use frames of 'playing God'.
- Intelligent design. Anti-evolutionists promoted frames around 'scientific uncertainty' and 'teach-the-controversy'.[14]

The crux of the issue

Matthew Nisbet has put his finger on the crux of the issue, in stating that frames can be used effectively in conjunction with segmentation studies to better know what messages will appeal best to what types of audiences. This can provide, he says:

> ... a deductive set of mental boxes and interpretive storylines that can be used to bring diverse audiences together on common ground, shape personal behavior, or mobilize collective action.[11]

He has also said that while we spend a lot of effort on messages in the media or advertising, we should not overlook interpersonal sources of information, as the frames used by influential peers can be incredibly strong.[15]

Social media as a propagator of frames

Social media has had a very distinct role in how frames can be analysed, as they allow researchers (for good or less good) to monitor social media and see which frames (or memes) are proving popular with the public, and then to use these to their own advantage (discounting LOL cat memes, of course).

The mainstream media, which has generally been agenda-setting, now often uses the frames that arise out of social media to frame their own stories – particularly stories of community outrage or concern. This leads to the media mirroring the way that people have responded to a story, and giving the public back a story that aligns with their existing views on it.

This can present challenges for science communicators, however, as often several frames have to be crafted to reach distinct audiences and when one crosses into the mainstream media, or gets widespread shares on social media platforms like Twitter or Facebook, it often becomes the dominant frame – which might not be the frame you want to reach all people. For instance, if you want to frame a message on climate change to a conservative pro-development audience about the investment growth from renewable energies, that is probably not the frame you want to get out to pro-environment groups.

Framing super powers

The ethics of framing

Which inevitably brings us back to the ethics of framing. While it is not quite a super power that could be used for good or for evil, it can be quite powerful, and can used by the forces of good or not so good (aka GONSG). Yet there do tend to be checks and balances built into most people's heads in the form of subtle personal bullshit detectors. If scientists or science communicators step over the line too often, and start engaging in hyperbole, then those bullshit detectors do have a habit of going off like the end of school bell. And if that happens too often, trust and credibility in you will flee as fast as a class full of primary school children on the last Friday before summer holidays.

What to do with what you now know

Framing is a very influential technique of giving certain information priority over other information. So when you are next working with an audience and have to explain something complex, have a think about what they might most want to hear (not what you most want to tell them) and the messaging and metaphors by which they might most understand it.

For example, my father never believed much in climate change. He knew the climate was changing, sure, but climate change was all about accepting that he would have to change lots of his behaviours like having a big car and a big house and questioning his values around the importance of economic growth. Each Christmas we'd end up arguing about climate change things until I did some research on framing. Then I said to him, 'Here's some information saying the big investment opportunities for the future are going to be in sustainable industries.' Suddenly he was interested. And then we ended up having a big talk about sustainable industries – that even edged into climate change territory without using the words climate change.

He'd been framed!

If you are going to engage in framing, don't drift over to the dark side as it will invariably bring you undone. If you have learned anything at all from Darth Vader, you should have at least learned that!

Researchers Druckman and Lupia have defined the challenge:

> If science communicators can choose frames that draw prospective learners' attention, while staying true to the actual conduct and principles of the underlying research, they can provide great value to audiences.[16]

Key summary points

- Framing is about giving some information about a topic precedence over other information.

- Framing information in different ways can increase or decrease the way it is accepted by an audience, based on how well it aligns with people's own values.

- Contentious issues have many different interest groups competing to have their own preferred frame as the dominant one.

- Framing different messages on the same topic to different audiences, without mixing them, can be about as tricky as it sounds.

- Framing can be a very powerful communication tool, but if misused or exaggerated can lead to a loss of trust.

13

Who's afraid of public speaking?

'According to most studies, people's number one fear is public speaking. Number two is death. Death is number two. Does that sound right? This means to the average person, if you go to a funeral, you're better off in the casket than doing the eulogy.'

– Jerry Seinfeld, comedian

Fear of public speaking – or glossophobia – is regularly listed as one of the greatest fears most people have, ranking right up there with a fear of death or a fear of being abducted by aliens who conduct invasive research through unmentionable parts of your body. Okay, not really that last one. But according to a very highly-cited Gallup Poll conducted in the US in 1998 and 2001, the only thing people actually fear more than public speaking is snakes.[1]

However, despite this statistic being widely cited, because after all it makes a great story, the poll findings have rarely been replicated, and more recent polls list things like terrorism and corruption among as our highest fears. A 2017 study by Chapman University found the biggest fears amongst US citizens were corruptions of government officials, healthcare, pollution and not having enough money for the future. A fear of public speaking came in at number 52 on the list of 80 fears, being beaten by things like being hit by a drunk driver or identity theft[2] (see Fig. 13.1).

In researching people's top fears, I found that a lot of the surveys in Australia and the UK were conducted by popular culture websites, favouring what is 'hot' in the media at the moment, rather than actual deep and persisting fears (like a fear of discovering you are quoting data from a dubious online poll!).

In fact, to really get to the bottom of the debate over whether public speaking was feared more than death, researchers analysed college students in the US and found that while public speaking was selected more often as a common fear among a list of fears, including death, when they were asked to select a top fear, death was selected most often.[3]

So yes, we generally have a fear of public speaking – but it is not as bad as is sometimes made out.

Another researcher has postulated that our fear of public speaking comes from our primitive desire not to stand out from a group where we would be more likely to be eaten by predators.[4] Is that comforting in any way to know that when you are up on the podium, it is not a fear of the audience that your brain is trying to tell you to run away from, it is a fear of lions and tigers and bears?

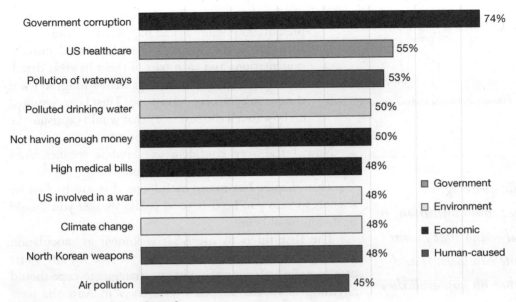

Fig. 13.1. Top 10 fears of 2018[2]

But like it or not, anyone working in science or science communication has to give public talks. Lots and lots of them. Sometimes just to a small room of colleagues, but sometimes to a huge giant gathering the size of a rock concert audience.

So let's first acknowledge that public speaking is not easy. Nor is successfully communicating anything through spoken language. For starters, effectively understanding your audience is hard, and then you need to convey complicated information to them that you need to understand fully yourself. And you need to translate it effectively into much simpler language that is accessible to your entire audience.[5]

It can be a very big ask.

However, once again science comes to the answer, with some recommendations for the best public speaking strategies, backed by some actual research.

Let's look at the big fear issues first. How to overcome the anxiety of speaking in public. According to Beverley Flemington, writing in *Psychology Today*, most fears of public speaking come from fears of being judged, rejected or humiliated by an audience (rather than being eaten by them).[6] But you can get so scared of being judged by others, that you forget that you are the one with the power to control how others perceive you. It is a matter of tapping into those powers and moving from feeling helpless to feeling in control.

Trust me on this. I used to go nearly catatonic with panic before giving a public talk. My body would start sweating and shaking and I'd have to rush off to the toilet, and I'd feel my heart going crazy in my chest like it was going to pop. But I have to do a lot of public speaking in my work, so I started looking at the research into what drives panic attacks and how to counter them.

The sum of all fears: a dead snake speaking in public!

Public speaking, like science communication, is a skill – and that means anyone can learn it and improve on how well they do it with time and practice.

I then started visualising a successful presentation before it had happened and learned to float above the anxiety. Then I planned out the presentations and each part of them in great detail, including what to do if things went wrong, so I was the one in control of everything. Then I concentrated on how to tell the best story that would captivate the audience (as outlined in Chapter 8).

Now I feel very confident as a public speaker. And seriously, if I can do it, you can do it too.

This chapter has many useful tips that are backed by science, to help you be that good public speaker you should be dreaming of being.

The first tip is to use what is known as inoculation training. Put simply this is preparing yourself for any worst case scenario by having thought out strategies to cope should anything untoward happen. Researchers in Australia have found that comparing the fear with the actual chances of the fear being played out (like the fear of forgetting every word you have to say, or a fear of fainting from anxiety) can help you understand that the fear is way out of proportion to the probability of it ever happening. Then, when you've walked yourself through your best possible responses (like having notes if you forget what to say, or just the fact that you've never seen anyone faint at a lectern before) the effect of the fear is minimised.[7]

What makes a good presentation?

Researchers at Quantified Communications, based in Austin, Texas, have analysed over 100 000 presentations to find the formula for a good presentation.

Maegan Stephens, who works for the company, has said that you don't need to be born a great public speaker to give a great public speech, and you can achieve the skills you need with a bit of dedication, hard work and practice. The team at Quantified Communications analysed presentations from politicians, executives and other keynote speakers, and developed some algorithms that define a good speech.[9]

They also measured facial expressions, gestures and vocal cues, and ranked speeches on factors such as clarity, trustworthiness, confidence and warmth. All together they use 25 different metrics to evaluate a speech and give seven basic tips to work on if you want to improve your public speaking:

1. **The audience is the most important 'person'**
   Know what your audience needs to know, and what constrains them from knowing it – such as background knowledge or time available.
2. **Content is really a priority**
   Without good content you don't really have a great public speech – no matter how well you talk (politicians *please* take note). And don't imagine that the way

Dealing with a fear of public speaking

The US Mayo Clinic has put together a checklist to help overcome any fear of public speaking you might have. They acknowledge that such fears can range from slight nervousness to paralysing fear and panic. With preparation and persistence, they assure you, it is possible to overcome any fear.

- **Know your topic.** Really know your topic so you won't get lost or get off track, and if you do you can recover quickly.
- **Get organised.** The more you plan everything the more in control you will be and the less nervous you will get. Use an outline card or notes to help if you need it.
- **Practice, and then practice some more.** The mirror is a good place to practice, but doesn't give great feedback. Friends and family can be better – but only if they are prepared to be honest.
- **Challenge specific worries.** List your specific worries, then directly challenge them by identifying probable and alternative outcomes or the likelihood that your feared outcomes will actually happen.
- **Visualise your success.** Imagine how well your presentation or talk will go. Positive thoughts can help decrease some of your anxiety.
- **Do some deep breathing.** This can be very calming, so remember to take two or more deep, slow breaths before you get up to the podium and also during your speech.
- **Focus on your material, not on your audience.** Put your energies and your concentration on where they need to be, not where they don't need to be.
- **Don't fear a moment of silence.** If your mind goes blank, it may seem like you've been silent for an eternity, but it is probably only a few seconds. Use the time to take a sip of water or just take a few slow deep breaths and then resume again.
- **If you don't know the answer, speculate.** Never make up an answer that you don't know, rather speculate what it might be.
- **Recognise your success.** Don't neglect to give yourself a pat on the back after you are done, even if you weren't as perfect as you had hoped.
- **Get support.** If you really need to, then join a group that offers support for people who have difficulty with public speaking, such as Toastmasters.[8]

you write, or the way that people read is the same as the way people speak and listen.

I'm sure we've all sat through a conference where somebody has endeavoured to read out their research paper and the audience has been sitting there with matchsticks propping up their eyelids thinking, just email me the paper and get onto the next speaker!

3. **Know the theories of primacy and recency**

The theories of primacy and recency state that your audience is most likely to recall the very first things you say and also the last things you say. Which tells us that if you start with a long list of apologies and then end with a long list of thank yous, they will be the most remembered bits of your talk. The middle bits, where you actually give all the really important stuff – not so much.

Also, the introduction is often the most important part of any speech, not only because people remember the first things you have to say, but they also tend to form an impression on of you in the first 15 seconds or so.

So some sage advice is don't start your speech with, 'Hi, my name is Craig and I'm here today to talk to you about science communication.' There is a high chance you have just been introduced in exactly those terms. Better to start with something snappy and memorable, like, 'I guarantee that 80% of you are going to forget everything I tell you today except the first things I say, and the last things I say.'

4. **Build in rehearsal time**

Musicians practice. Athletes practice. Doctors practice (sorry, bad joke). But if they practice to become good at what they do, then logic dictates that if you want to be good then you should practice too.

I find the best time to practice speeches is actually lying in bed, not standing in the lounge room practicing to the dogs. I run over things in my head that I am going to say, and picture myself in the environment where I am going to say it, also making myself more comfortable in my mind with that environment.

However, you might be one of the many people who benefits more from practicing standing up in a room with people who can give you some good feedback. Practice and train your body and mind to do what you need them to do in a situation that is close to that you are going to actually going to be doing. Experiment a bit and find the thing that works best for you.

5. **Try to maintain a natural, authentic speaking style**

This is easier said than done, of course, but it gets easier with practice, and it also gets easier if you focus on one person in the audience and think in your mind that you are talking just to them (without staring just at them and creeping them out of course).

Quantified Communications rate a speech's delivery as being more important than its content, as this is what the audience are more responding to.

6. **Get a recording**

If you are still feeling unsure of yourself, have someone record you speaking and then sit down and analyse it. It can be an ouch moment though, as we are all hyper-critical of our own performances, but you can more easily see what you are doing right and what you are doing wrong.

7. **Track your performance**

Most speakers use the 'that felt good' measure to gauge whether a speech was successful. But realistically, it is about as effective as telling yourself that breaking wind in the work place felt good – and not checking what anybody else felt about it.

What makes a good presenter?

We all know there are some great public speakers out there and some other people who should have stamped on their files 'Never to be let loose in front of an audience'. And yet some great orators have admitted they were very, very nervous at public speaking – including Mahatma Gandhi and Abraham Lincoln.[10]

So the good news is that confident public speaking can certainly be learned.

According to Harvard Business professor, John Antonakis, a cornerstone to being a better public speaker is adopting charismatic practices. His research lists 12 charismatic leadership tactics – or what he calls CLTs – that are used by most leaders. And, he claims, that when executives use these tactics, their leadership ratings have been shown to rise by up to 60%.[11]

Now admittedly you are not trying to lead a large corporation – you just want to get through giving a good public talk without fainting or falling over, and you also want to make a good impression and actually pass some relevant information to your audience. So let's look at the key leadership tactics that apply to public speaking.

- **Connect, compare and contrast**
  According to John Antonakis, professor of organisational behaviour at the University of Lausanne in Switzerland, charismatic speakers help listeners understand, relate to, and remember a message through effective use of metaphors, similes, and analogies (see Chapter 7) that enable you to connect to your audience.

- **Stories and anecdotes**
  Both have narrative power and should make a point, by comparing the situation you are talking about to one in your story. And the more personal you can make the story, the better.

- **Contrasts**
  These are a rhetorical device best summed by John F. Kennedy's 'Ask not what your country can do for you—ask what you can do for your country'. A good contrast states your position by contrasting it with an opposite position.
  Another contrast that Antonakis quotes is the line from a business manager to his staff: 'It seems to me that you're playing too much defense when you need to be playing more offense.'[11]

- **Engage and distil**
  Rhetorical questions and three-part lists are both very effective ways to engage and distil a message, such as, 'So you probably have three key questions today: What am I going to tell you that I will talk about? What will I then tell you? And then what will I reiterate that I have told you?'

- **Show confidence and passion**
  You probably really care about your own work, or whatever it is that you have been

90 years ago, rounded up to the closest factor of ten, our genetic predecessors conceived, though not in a biological sense, a new independent geographical and political entity...

President Lincoln has second thoughts about letting his science advisor write the Gettysburg Address

asked to talk about – so show that in your talk. Let the audience feel how important your work is.

Randy Olson, the author of *Don't be Such a Scientist*, says you need to connect with the audience via four of your organs: your head, your heart, your gut and your sex organs. He says the object is to move the process down out of your head, into your heart with sincerity, into your gut with humour, and then ideally into your sex organs – if you can – with sex appeal.[12]

If you are not sure what that last one means, go online and watch Brian Cox in action. Or Carin Bondar. Or Isabella Rossellini. And if it still doesn't make sense – just skip that bit of the advice.

Putting it into practice

These are all useful hints, but what should a good speech actually look like in structure? (See Chapter 8 on storytelling.) We know it needs a beginning, a middle and an end, but what else? Well, once again research comes with the answer. Nancy Duarte, who runs her own company providing presentation services, has analysed the structure of many, many influential speeches and found a common structure to them, as is shown in Fig. 13.2.

Putting the structure into plain English:

1. Start by explaining 'what is' or what's the problem or topic you want to address, or to change.
2. Then explain 'what could be' which is what an ideal outcome could be.
3. Next explain what needs to happen to reach the goal, or the processes you have gone through.
4. When you've explained all of your points, end the speech with a clear vision of what the audience could expect.

This is very similar to the ABT approach outlined in Chapter 8, and if you are a TED talk fan, look up Nancy Duarte's TED talk on speech theory.

The importance of gestures

And speaking of TED talks, another study looked at TED talkers with the volume on and the volume off, and basically the speakers all got the same rating with or without sound. How can that be?[14]

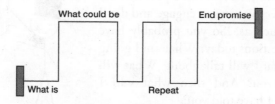

Nancy Duarte's structure of a great speech

Fig. 13.2. Nancy Duarte's Sparkline structure[13]

Well, yes, you might say that is because all TED talks sound a little bit the same, but there is much more to it than that. We tend to judge a speaker largely on how they look and stand and so on. Hand gestures, smiling and vocal variety are very important for captivating an audience (and yes, there is even a TED talk on this too).

Adopting a power stance

Vanessa van Edwards, of Science of People, found that the most popular TED talkers used an average of 465 hand gestures, while the least popular talkers used only 272 hand gestures.[14] Regardless of what they were saying.

Hand gestures can be used to capture your audience visually and to convey emotions and really emphasise what you are saying. Spencer Kelly, who has been an associate professor of psychology at the Center for Language and Brain at Colgate University in New York, found that gestures make people pay attention more to the acoustics of speech. According to Kelly, 'Gestures are not merely add-ons to language – they may actually be a fundamental part of it.'[15]

There is even a great list of hand gestures on Wikipedia, under 'Hand gestures' that give some examples of how they are used – such as the Clinton Thumb, a closed first version of the Trump Thumb and Forefinger. But for a guide to the best hand gestures, Vanessa Van Edwards of Science of People recommends:

- using your fingers to count off a list (and the ideal list is three, because we just love lists of three)
- using your fingers or hands to show something is big or small
- the rising palm – hold your palm out like you are asking for a handout, and rise it towards you. It draws the audience towards you, as if what you say is going to be more important
- the pointing finger – which can be used to indicate to the audience, or to emphasise a point
- hands towards self – which can be one hand placed on your breast, or two hands sweeping towards yourself – which emphasises something that is about you or more personal
- the power stand – legs apart, chest slightly forward, maybe hands on hips. Stand like a winner, like an expert. You will feel like an expert and a winner and more likely be seen as one.[14]

There are many, many more articles and research findings that have other lists of key things to do (and not do), but here is a short summary of some of the best advice:

- Don't remain rigid, hiding behind a lectern. Move around freely, but don't pace. Own the available space on the stage.
- Don't fiddle with anything. Keep your hands free for expressive gesticulation. (However, don't end up doing weird repetitive things! Watch a video of yourself to make sure.)

- Channel your inner Patrick Stewart (Captain Jean-Luc Picard from *Star Trek: The Next Generation* for those not up to date with cultural references), and try and use the five Ps of good speaking:
  - projection – get your voice to carry across the room
  - pace – use a variety of talking speeds to suit what you are saying and try and avoid the rapid robot talk of the nervous
  - pitch – your voice should rise and fall a bit to emphasise thing (watch politicians and TED talks and note how they do it)
  - pronunciation – speak clearly and get tricky words right, otherwise it distracts from what you want to say
  - pauses – stop and take a breath every now and then; pause for emphasis or after having made a strong point that the audience might need to consider a little bit.
- Try to enjoy yourself (or at least look like you are enjoying yourself!).[16]
- Repeat yourself to increase retention among your audience; however, don't overdo it. No, don't overdo it. Definitely don't overdo it.
- Establish yourself as an authority on the subject which will enhance your credibility.
- Use humour only if you are good at it (and again, like the fart in the office, don't use your own judgment on how funny you are – ask someone who you know will tell you the truth).
- Smile (without being creepy – we all know someone who should have stamped on their file 'Do not smile in public!').

When things go really, really wrong

Well we already know that's not actually going to happen to you, right? But it is a good idea to have a strategy in your head to cover off the idea of it, okay?

Professional speaker Scott Berkun recommends if you find something had gone terribly wrong – like the IT has failed you, or you are hopelessly lost in your notes – that you should look for a face in the crowd that seems supportive. He says that can become an emotional base. Look to that person for support, and then use that to keep you moving forward and get back on track.

Death by PowerPoint

And then there is PowerPoint. There is always, always PowerPoint. A tool that when used well can be a great boon to communication – but when used poorly can become a real problem for you.

I know it is considered a thing to bemoan the use of PowerPoint, as if it is the eighth deadly sin, but if used creatively and well it really can assist your communication efforts.

I attended a great presentation once where the PowerPoint slides did not contain a single word – each slide had a picture from a Charlie Chaplin movie on it, but one that was integral to the message that the speaker was giving. It was a fantastic presentation, and really stood out from the others.

Science blogger Chad Orzel, in a blog post on 'The quirks of scientific public speaking', raises the very strange relationship many of us have with our PowerPoint slides. He said that professional scientists spend a great deal of time speaking in public, but frequently appear sort of amateurish compared to speakers in politics and business because of poor use of slides.[17]

Scott Berkun lists six reasons why he feels speakers use bad slides:

1. Bad slides are less work. Nearly everything in the world we know to be stupid is easier to do than doing the right thing. For unpleasant things, most people are happy to do just enough not to be horrible and move on.	**2. The slide is used for more than the presentation alone.** An example of this is the slide deck that is meant to stand alone as notes when circulated.
3. Organisational tradition demands bad slides. If most people at an event, or in an organisation, do the same old boring, hard-to-understand, ugly distracting template, bulleted list wall of death, then that becomes the standard.	**4. It makes it look like you've done tons of work.** To the average eye, a dense, heavy, slide deck can look like more work has been done – when in reality a simple, clear, concise slide deck requires much more work.
5. They really don't know there's an alternative. Some people don't get out much and don't see really good presentations.	**6. Speakers use slides as their own notes.** Slides should be for the audience, not the speaker, and while it's okay to leave cues for yourself in slides, they must be minimal enough that they don't ruin the audiences' experience.[18]

On the other hand, UK researcher Chris Atherton found that the less data on the slide you use, the more information the brain can take in. So rather than crowding dot points onto a slide, only give the barest essential cut down text (maybe only a single phrase), ideally with an illustration or graph to support it visually. He has said that sparse slides lead to fewer competing attention demands, and sparse visual cues lead to better encoding of information.[19]

In short, more slides with fewer words on them, is preferred to few slides with more words on them. And visual supports – whether it be a graph or a still from a Charlie Chaplin film.

Appropriately, he has put his research into a PowerPoint on Slideshare. Check it out.[19]

Data visualisation

There has been a huge growth in data visualisation but, like giving a user too many fonts to play with, it has led to the good, the bad and the ugly – with many infographics, for instance, being high on graphics, but low on info. Scientific data visualisation at its best is a mix of science and art and there are a growing number of designers and artists who excel at this – largely by not using the Excel program.

If you look at some of the earliest instances of effective data visualisations, you will see how the graphics were used to show the data in a new and compelling way. In 1854, a London epidemiologist named John Snow (clearly not the Game of Thrones' 'you know nothing, Jon Snow!') showed that a cholera epidemic that killed over 500 people was based around the closeness to a neighbourhood water pump. He used a map of the city and plotted every death as a bar chart according to where they lived, visually demonstrating the closer to the pump, the more deaths occurred. This in turn led to better sewage in the city.[20]

And in 1858, Florence Nightingale used an innovative form of a pie chart – the rose diagram – (or 'wedges' as she called it) to visually demonstrate the major causes of mortality for soldiers in the Crimean War (see Fig. 13.3). By shading the segments to show death in combat versus other causes of death it was easy to see there was an issue that needed to be addressed – although you can also show that she had to explicitly explain what the diagram meant in a side text as her audience was not familiar with seeing data in this form. (A lesson to consider if you are doing something unfamiliar!)

Interestingly, her research was a part of a Royal Commission into the causes of mortality of soldiers in the Crimean War, and she worked alongside William Farr, who was quite a pioneer in statistics, but who did not favour including the visualisation in the report. It was only included upon Florence Nightingale's insistence.[21]

Yet we should also look beyond the good, for history has more than its fair share of data visualisation (a 'viz' for those hip enough to say viz) that are bad or ugly. They include 3-D pie charts that distort the relativity of slices, over-cluttered 3-D bar charts, and standard pie charts whose segments don't add up to 100% (like an often-quoted Fox News pie chart of the 2012 US Presidential candidates for the Republican party that showed 70% support for Sarah Palin, 60% support for Mitt Romney and 63% support for Mike Huckabee – adding up to 193%!).[22]

Other common errors include:

- wonky data
- wrong choice of visualisation

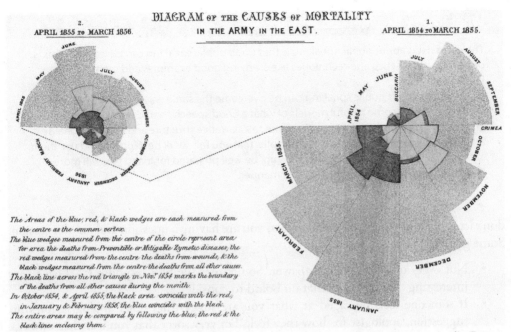

Fig. 13.3. Florence Nightingale's diagram of the causes of mortality in the army in the East, 1858

- too much information
- inconsistent scales
- cropped axes
- missing labels or annotations.[23]

Talking to a hostile audience

There are also times in your career when you might find yourself talking to a hostile audience – whether explaining a contentious science finding or an unpopular scientific decision. Before you draw the analogy that these could be times when that primitive fear of being attacked by predators might really come true, there are many things you can do to work with such an audience rather than against them. For instance:

- Listen to your audience before you talk to them.
- Get out from behind a desk or podium and sit or stand closer to the audience.
- Dress in a way that does not distance you but shows you are more like your audience.
- Don't bet bogged down in details of the science if their concerns are about something other than the science itself – such as who gets the benefits and who gets the risks.
- Talk in terms of human impacts and human benefits.

There are times though when you need to step away from the inclusive strategies, such as when questions and answers get out of control. Let people have their say, but

What to do with what you now know

The good news about public speaking is that good public speaking can be learned. This is done through practice and techniques like studying good examples and mental visualisation.

Anxieties about public speaking can be overcome the same way.

There are also some helpful models of what a good speech should look like in structure and content that you can draw on. Give your talk narrative structure and flow. Personalise it and use anecdotes. Look for the key principles that you feel work best for you and use them.

So when your next big speech comes up, be well prepared for it and you will more likely nail it in a confident and successful manner.

don't let them take over, particularly when you are having a question and answer session. Some good guidance includes:

- Not every question has an answer. Sometimes the best response is, 'That's very interesting. We will take that on board to consider.'
- If someone feels outraged at what you are suggesting, don't apologise for your suggestion, apologise for how they feel. 'I'm very sorry that you feel that way.'
- Not every question is a question. Many will be statements, which you can choose to respond to or simply thank the person for their comment and move on to the next question.
- If there is a particular individual who wants the platform, you are better off going into the audience to be near them than letting them come to you. The closeness changes the relationship too and makes the hostility harder to maintain. (And don't pass over your microphone. Hold it close to them but keep your hold and your control of it.)

Key summary points

- Feeling anxious about giving a public talk is normal, but you can overcome anxiety with practice and by understanding what drives your anxiety.

- Good public speakers use the characteristics of charismatic leaders, which can be learned.

- A good presentation can be modelled upon the structure of other good presentations.

- PowerPoint can be effective – if used well.

- When dealing with a hostile audience, use risk communication strategies.

14

Getting engaged

'Engage!'

- Captain Jean-Luc Picard, *Star Trek: The Next Generation*

The more I have conversations with people about community engagement, the more I feel that most people don't actually know what it means, and they really just like the sound of it. Or they have their own meaning for it.

Community engagement, or public engagement, has become a bit of a jargon buzzword for many types of science communication activity. And often, if you run the phrase through Google translator using the bullshit setting, you will get results similar to those in Table 14.1.

What is community engagement then?

It is a term for a particular type of engagement that, in short, is about sharing information to make decisions. And to help you understand that, we are going to look at how it evolved and what it is today – like you might analyse a breed of parrot or spider, by examining its evolution to what it is now.

The US Centers for Disease Control has developed a comprehensive document called *Principles of Community Engagement* that addresses much of the theory and practice of community engagement. It has a good working definition of community engagement that captures its key features:

> ... the process of working collaboratively with and through groups of people affiliated by geographic proximity, special interest, or similar situations to address issues affecting the well-being of those people. It is a powerful vehicle for bringing about environmental and behavioural changes that will improve the health of the community and its members. It often involves partnerships and coalitions that help mobilize resources and influence systems, change relationships among partners, and serve as catalysts for changing policies, programs, and practices.[1]

It's a bit unwieldy for explaining things to your grandparents though, or trying to use as a pick-up line. A simpler definition, provided in the South Australian Government's *Community Engagement Handbook,* is:

> Community engagement is about involving the community in decision making processes, which is critical in the successful development of acceptable policies and decisions in government, the private sector and the community.[2]

Table 14.1. Translating engagement objectives to what was actually meant

Original	Translated to what was actually meant
'We need more community engagement with our target audiences.'	'We need bigger audiences.'
'We want to undertake more public engagement'	'We want to tell people what we are doing.'
'There is not sufficient community engagement on our programs.'	'People don't have a clue what we're doing.'

So – are we clear on that? It is about INVOLVING the COMMUNITY in DECISION MAKING.

A good example of engaging with communities, rather than trying to sell them something, was a study of best practice community engagement relating to windmills in Australia.[3] The initial problem was that many communities were very divided on the benefits of having windmills built in their areas. In general, those that were getting money from leasing land for the construction of these huge white windmills were in favour, and those that weren't getting a benefit were worried about their impact on the landscape and possible health impacts from low frequency noise generated by the turbines.

The study found that effective engagement needed to go beyond the standard consultation methods traditionally used, where a group of experts stand up and tell the community why it will be good for them, giving them some science and statistics and then packing up and going home. That works for a community that is totally on board with what you are doing – but such communities are rare more often than not, and you are likely to just stir up more concerns.

Good engagement, the report recommends, is based on fostering relationships, building trust, growing feelings of ownership and instilling a sense of collaboration through provision of 'meaningful, ongoing opportunities for the community to participate in a project's development'.[3]

The study defined several key principles for best practice:

- mutual benefit
- mutual respect
- relationship building
- authenticity
- appropriateness
- ongoing engagement
- transparency
- responsiveness.

The study also mapped out the different activities that would align with different levels of engagement (Table 14.2).

This is based on the Infographic Bible of Engagement, or IAP2 Spectrum of Public Participation. Developed by the International Association of Public Participation

(IAP2), it is something that everybody who is serious about understanding public engagement should know, and divides engagement into five key levels, ranging from simply informing right through to empowering:

1. inform
2. consult
3. involve
4. collaborate
5. empower.[4]

Each level has increasingly more engagement from straight information provision, obtaining feedback, working closely together, partnering and finally handing over decision-making power to the audience. Techniques that suit each level of engagement are given in Table 14.3.

This is such an important model that I highly recommend getting it tattooed onto your torso like I have done. (Okay, I haven't really, but you knew never to accept random claims as facts just because they sound good, right?) And one of the biggest fall-downs with engagement is when one model is promised – such as collaborating – but another is actually delivered – such as involving – because there will be a gap between expectation and delivery.

Many community groups or individuals might come to an engagement with a preference for a higher level of engagement, such as collaboration or empowerment, but have to be told that is not going to happen, and all they are going to get is involvement. It is important to understand if there are differences in what is wanted and what you are planning to deliver, so mismatched expectations can be addressed early.

As there are many different models of community engagement to draw upon, it is very important to select the model most suited to the audience/s you are working with. For instance, community members that are largely resistant to your messages, with low awareness of the science, are best engaged with by an involve model. Those with predominantly high support for, and moderate awareness of, the science, are best engaged with through a collaborate model.

Nobody loves mice

Ceding control over decision making, or even sharing decision making, does not come easily to many organisations, who might fear that the public will make a wrong decision. But evidence shows this is rarely the case, and people tend to make sensible decisions when fully participating in the decision-making process with access to all available information.

Here's a story to illustrate this. I was at a seminar some years back and a researcher told us about the development of a theoretical virus to make mice sterile and prevent mice plagues. The story goes that the researchers went around rural Australian communities and asked what people thought about them developing a virus that might make mice sterile. The response was not so great, with a lot of concern about whether it might cross into native mice, or into the human population and so on.

Table 14.2. Activities that would align with different levels of engagement[3]

	Standard development approach with compliance-level community engagement only. Doing development 'to' a community		High level of engagement or community–developer partnership. Doing development 'with' a community		Community-initiated wind farm. Doing a development 'for' and/or 'by' a community
	Inform	Consult	Involve	Collaborate	Empower
Community engagement objective	Provide balanced and objective information. Assist community in understanding the problems, alternatives and/or solutions	Obtain feedback on plans, options and/or decisions	Work directly with community throughout the process, from feasibility through to operations and decommissioning. Ensure concerns and aspirations are consistently understood and considered	Partner with community in each aspect of planning, development and decision making, including the development of alternatives and the identification of the preferred solution	For the community to lead the development of the wind farm. Place final decision making in the hands of the community
Promise to community	To keep the community informed	To keep the community informed. Listen and acknowledge concerns. Provide feedback on how community input influenced the decision	To work with community to ensure concerns and aspirations are directly reflected in the alternatives developed. Provide feedback on how community input influenced the decision	Look to community for direct advice and innovation in formulating solutions. Incorporate advice and recommendations into the decisions to the maximum extent possible	To implement what the community decides

Community engagement outcomes	Standard development approach with compliance-level community engagement only *Doing development 'to' a community*		High level of engagement or community– developer partnership *Doing development 'with' a community*		Community-initiated wind farm *Doing a development 'for' and/or 'by' a community*
	Inform	Consult	Involve	Collaborate	Empower
	Securing landowner sites Gaining planning permission Meeting compliance	Minimising objections Effectively managing complaints Good landlord relations A level of community trust	Long-term broad local social acceptance of the wind farm Strengthened local relationships and trust Local advocates	Community– developer partnership, where a portion of the wind farm is owned by a local community Greater community benefit A welcoming and supportive host community Strong local relationships and trust Timely development of the wind farm Long-term social acceptance and local advocates for the wind farm	Benefit sharing model tailored to the local context Community-scaled project Harness the skills and capital of the community Upskill community members to manage the project Largely community owned and controlled

Table 14.3. Techniques for different levels of engagement

INFORM	CONSULT	INVOLVE	COLLABORATE	EMPOWER
Fact sheet Website Advertisement Media release Newsletter Email list Social media	Public comment Focus groups Surveys Stakeholder meetings Shopfronts Phone hotlines Briefings Feedback forms Social media	Workshops Deliberate polling Social media Advisory groups Discussion forums	Advisory groups Discussion forums	(Depends less on technique and more on relationship) Summits Deliberative democracy Citizen science Deliberative forums Citizen juries

The researchers thought about this and then, instead of just consulting further, undertook a fuller community engagement program. They went back to the communities who were greatly affected by seasonal mice plagues and talked with them. Their starting point was an agreement that nobody loved mice, so common ground was achieved to begin with.

Working with the communities, they then went through all the options for getting rid of the mice, such as … well maybe importing lots of cats, or shooting them, or poisoning them, or hiring a Pied Piper perhaps – I don't know for sure, but I do know that one of the options included developing a virus that might make mice sterile.

When the members of the communities were actively taking part in the decisions about how to control mice, and assessing the risks and benefits and costs and feasibility of each option, they invariably found that the virus was the best option.

The research project itself never went ahead, for a complexity of reasons, but these days, similar discussions need to be had about controlling pests with gene drives. The moral though is: share data and decision-making processes and the public generally make sensible decisions.

The evolution of community engagement

Many researchers have mapped out the evolution of community engagement and the many disciplines and interest areas that it sprang from. A key trigger was the UK Government's House of Lords Select Committee on Science and Technology in 2000.

The House of Lords, the upper house of the British Parliament, had held an inquiry into the relationship between science and society, and their report marked a significant turning point in science communication worldwide. A key finding – which still hasn't penetrated as widely as it should – was that educational activities were no longer enough to engage the more sceptical and less-engaged members of the public. The report recommended a different approach – through dialogue, or engagements – whereby those seeking to promote science or science-based evidence, listened to the concerns of the public.[5]

This led to a worldwide flurry to incorporate more dialogue-based methods into public consultations relating to science (especially for contentious issues such as genetically modified foods or human embryo research – which were the two hot topics of the day). These public consultation activities trialled many different methods such as debates, public meetings, workshops, citizens' juries and consensus conferences.

Concepts of best practice community engagement have continued to evolve, based on the growing field of theory and practice that has largely been developed in Europe, often around contentious technologies. As an example, the Rathenau Institute in the Netherlands developed a best practice guide for community engagement that included:

- differentiate between the risk issue and the broader debate, and actively address the risk issue
- involve key NGOs in developing policy and assist the involvement of smaller NGOs
- provide clear information about the risk governance and uncertainties
- create a public agenda that enjoys wide support
- inform the public about the societal aspects of the topic
- give citizens a voice by means of small-scale engagement activities.

One of the findings of the Institute was also that small-scale engagements could be as effective as the larger-scale, high-profile engagements that governments had supported, including the NanoJury and GM Nation that were held in the UK.[6]

I am sometimes asked for my opinion on running a citizen's jury by someone who is clearly captured by the idea of it. And my general advice is that of a lawyer – you can never tell quite which way a jury will go. It can be a high-risk strategy, which can be lessened by holding more smaller juries rather than one big one.

The Rathenau Institute also found that rather than a 'one-size-fits-all' approach, different activities were more effective if tailored to the motivations of different segments of people.

More and more researchers and practitioners are coming up with different methodologies and practices and recommendations. This means that if you are serious about public engagement, there is an awful lot of research to draw on. But it also means that if you are serious about public engagement you might risk drowning in the published literature. One Australian researcher, John Gilbert, identified around 60 different types of programs just relating to engagement or education on bushfires.[7]

When communities need to own community engagement

Many engagements can be about empowering communities to make decisions about their own wellbeing, such as is done in relation to bushfires or environmental clean-ups. The purpose of the engagement is for the community to be an active partner in solving some problem.

When a community owns the engagement it also means that any trade-offs between 'winners' and 'losers' are made, and agreed upon, by the community. Don't try and do this for the community – you will end up as the bad guy. Not everyone will be able to feel that their concerns have been reflected in an outcome, as there will invariably be differences of opinion within any community – but there should at least be a majority of accord.

And for those who don't support an outcome, they should be offered ongoing conversations, and be kept informed of things.

Empowering communities to make decisions about things that impact them has benefits such as increasing the level of trust in your agency. Another benefit is an increased likelihood of people acting on risk mitigation information. A study of tsunami preparedness in a low-familiarity Alaskan community, for instance, found that empowerment increased community involvement and intentions to prepare for hazards.[8]

What makes good engagement?

Many experts have sat on mountain tops and mulled that question, and when the questing social scientist climbed up to them and asked what constitutes good community engagement, they have said, 'That depends entirely on the context.' And they are right. Any list of best practice activities you read will need to be adapted to your particular circumstances and objectives. But there are some principles that need to be understood.

As every community can be very different, and made up of groups of individuals who can also be very different, there can be no one formulaic approach that will work well across many communities and the type of discussions held in one community, might not work so well in another. This requires staff working in engagement to be skilled in various methods of community discussions, and to be empowered to experiment with them, adding to the learning base of knowledge.

Good engagement often begins with acknowledging and validating people's existing beliefs and attitudes, and then having conversations about what people value. This can give insights into how to best engage communities by using the topics of most value to them for framing conversations. It also allows you to better incorporate the values of the public into decision making and gauging public preferences in relation to key trade-off decisions.

So ideally, good public engagement for the development of a new technology would go something like this:

A scientist develops a new process or innovation, and before applying it he or she has a discussion with the community that will be most affected by it, as to how they would like the technology to be developed and used.

The community sit down with the developer, and the regulators, and other expert stakeholders on impacts and they discuss, in clear and reasoned ways, what types of applications should have resources put into them, and what types of products should be developed, assessing issues from all sides.

Then, with a firm understanding of the public's position, and perhaps even social licence to operate, the research is continued in a particular direction, capital for development is easier to acquire, and products are developed.

The public, the scientists, the funders, regulators and developers are all happy with the outcomes.

In reality, public engagement around the development of a new technology tends to go a little like this though:

A scientist develops a good idea and then hunts around for a use for that idea, focusing on areas most likely to attract development and commercialisation funding.

When the idea is eventually developed into an application it is taken to the market – where it succeeds or fails, for a variety of reasons.

If there is community backlash at that point, then engagement is undertaken to try and sell the benefits of the product or process, through explaining the science and its benefits and working with the public to minimise risk perception.

The technology stumbles along – though not everyone is happy about it.[9]

There are not many examples of the first model that spring to mind – and too many of the second, mostly based on the assumption that if an idea gets capital funding then it must be a good idea. But think how differently the path of GM crops might have played out if the developers had gone to the public before they were developed and said, 'Hey we have a new technology called gene technology, and this is basically how it works. So what would you prefer we do with it? And what types of applications would you most likely support?'

You can be pretty sure very few people would have said, 'Let's make a broadacre crop that is resistant to the pesticides or herbicides that your company makes.' More likely they would request healthier crops, or pharmaceuticals from crops grown in greenhouses. We'll never know though, for all the engagements on GM crops and foods we have today – and there have been a lot of them – have been conducted after the products were developed. But there are signs of better public engagement on the new uses of gene editing technologies.

What best practice looks like

The US Centers for Disease Control states that there should be three key outcomes from good community engagement:

1. **Capacity building:** Community empowerment theory stresses that no external entity should assume that it can bestow on a community the power to act in its own self-interest. Instead, those working to engage the community should, when

Different community engagement types may include:

Decision-based dialogues: These dialogues, designed in partnership with particular decision makers about specific decisions, will involve diverse perspectives, but the decision maker will undertake to take into account input received in making the decision.

Awareness raising dialogues: These might be held between particular decision makers and particular stakeholders, to both raise awareness about how decision making is done, and to inform and broaden the decision making.

Participatory technology assessments: Often called deliberative dialogues, these are held between decision makers and communities, and involve sharing research and analysis on issues, allowing the community to take part in setting the direction of the development of a particular technology, having looked at the pros and cons and current and future impacts.

Discursive public forums: These seek to generate opportunities for the public to take part in more intimate conversations, or group discussions, with experts and decision makers, to help inform better decision making.

Upstream engagement: The concept behind something being 'upstream' means very early in its development. Downstream engagement is after the science (or technology or product) is already developed. There has been a lot written about the need for good engagements to be upstream, but due to the nature of upstream being highly speculative sometimes, and no certainty of how a science or technologies trajectory will go, there is now talk of mid-stream engagement as perhaps being more realistic for many cases.

appropriate, offer tools and resources to help the community act in its own interest.

2. **Community empowerment:** Empowerment allows even marginalised individuals and groups to gain greater control over their lives and environment, and acquire valued knowledge and resources. Empowerment should be both a process and an outcome of community engagement.

3. **Coalition building:** Community engagement is increasingly acknowledged as a valuable process, not just in ensuring communities participate in decisions that affect them, but also to strengthen and enhance the relationship between communities and governments or other agencies. The concept is about public participation that facilitates engaging people in decision making at a local level.[1]

I could throw lists of best practice principles at you all day long (if only it were an Olympic sport, I'd be a contender for a medal), but it is more important to understand the underlying ideas behind each, for there is no one perfect model of engagement – as there is no one uniform set of perspectives within a community. Sometimes more informal methods of engagement are preferred, that might include single conversations. Sometimes larger engagement with many members of the community are needed.

I have taken part in many engagement activities on issues such as GM technologies and nanotechnologies and found that getting the model right for the audience is half the work. Good models for engaging with communities can be built upon:

- actively listening to people's attitudes, opinions and deeper causes of these opinions
- using very small steps to take people on a journey from their current position to a position where consensus can occur
- telling people what they already think and validate that it is reasonable to think that
- understanding and acknowledging people's values
- considering all views equally and giving consideration to all views
- insisting that different views are based on some supporting evidence
- making accurate and relevant data available to all participants.

It is also important to know that good community engagement requires cultural change within the organisation conducting it, as well as with the communities you are seeking to engage with. It is an ongoing process that is achieved over a significant period. For instance, the Western Australia Department Fire and Emergency Services Community Engagement Framework provides a five-year timeline to achieve best practice in engagement including continuous improvement.[10]

Multi-stakeholder engagement

Remember when options and what-to-do lists get a bit overwhelming, find the ones that resonate with how you work and stick with those. The others may or may not be used at some point in the future.

Another type of engagement is one where you need to engage with multiple stakeholders that represent a very broad range of stakeholder groups with competing views about a topic. They might also not get on particularly well together, and you end up with something like the United Nations trying to agree on a contentious topic. An example that springs to my mind – or memory – is an engagement on genetically modified crops that needed to involve farmers, organic growers, researchers, regulators, local authorities, community groups and scientists.

Let's just say they were not a group that you'd invite to a dinner party and expect to have a polite and genteel evening.

The intention of a multi-stakeholder process is to draw on a very wide range of knowledge and perspectives, and then to work though areas of differences, and hopefully arrive at a joint outcome. To achieve this, the many stakeholders must be willing to win on some things and lose on others. Any person who takes part in such a process and is not willing to concede any points at all is not really taking part in the process. Many of these types of engagement are about open listening and sharing perspectives, before moving on to any consensus seeking.

Science and Technology Engagement Pathway (STEP) model for community engagement

The STEP model was based on multi-stakeholder engagements around nanotechnology. It provides a good model for working on complex issues with complex audiences. The model was awarded a prize for best practice by the International Association of Public Participation, and is based on seven engagement principles:

1. **Commitment and integrity**
   A high level of commitment and integrity among organisers and participants, including mechanisms for transparency and accountability.
2. **Clarity of objectives and scope**
   Clarity concerning what the engagement is for, what's on the table, and what success would look like.
3. **Inclusiveness**
   Inclusiveness of the diversity of people and views, so that a range of perspectives is brought to the discussion and all those with an interest are able to be heard.
4. **Good process**
   Includes an appropriate and structured method; communication and consultation with participants throughout; and appropriate, independent oversight and evaluation.
5. **Quality information/knowledge sharing**
   Relevant, accurate, and balanced information and knowledge sharing.
6. **Dialogue and open discussion**
   Genuine, interactive, deliberative dialogue; opening up discussion rather than closing it down.
7. **Impact on decision making**
   Demonstrated influence on decision making.[11]

It should be up to the multi-stakeholder group themselves to trade pluses and minuses until some accord is reached by a consensus being negotiated. The process can be strengthened by having members of the general public in attendance to represent the wider public that many stakeholders will claim they are working in the interests of. This is very effective in reducing overblown arbitrary statements about the public's interest by activist groups, if the members of the public present don't agree with them.

Engaging the community through citizen science

There has been a growing interest in citizen science from both citizens and scientists, and the idea of it not being 'real' science is slowly disappearing as more scientists acknowledge both the benefits that can be gained from cooperating with citizen science projects, and that the data obtained can often be quite good.

Citizen science also provides a great opportunity for communicating science to a very engaged audience.

There have been lots of models and definitions of citizen science floated around, but the one I most like, from Scistarter, has four key principles:

Different models for community engagement

Researchers Rowe and Frewer found over 100 examples of engagement in practice, and it can make a significant difference to your outcomes as to which example is used. Yet observation shows models tend to be chosen to best suit the organisers' preferred outcomes, rather than for the participants' outcomes. Models they cited included:

- action planning
- citizens' juries
- community dinners
- computer-based techniques
- hotlines
- open houses
- study circles
- world cafes.[12]

Another study of different models of engagement found seven key types of models and looked at how representative they were and who they favoured. They were:

1. dialogues
2. participatory technology assessment
3. legal public hearings
4. consensus conferences with lay people and experts
5. extended consensus conferences that include interest groups
6. voting conferences
7. scenario workshops.

The analysis found that under most models, one group or another holds a key position of power:

- In the consensus conference, it is the lay persons.
- In the public hearing, it is the administrator.
- In participatory technology assessment models, it is often the scientific experts.

However, in two models – the voting conference and the scenario workshop – all participating groups enjoy equal rights, These models were deemed more 'balanced'.[13]

- anyone can participate
- participants all use the same protocol so data can be combined and be high quality
- data can help real scientists come to real conclusions
- a wide community of scientists and volunteers work together and share data to which the public, as well as scientists, have access to.[14]

Australian researchers Jenny Davis, Euan Ritchie, Jenny Martin and Sarah Maclagan, have said that at its best citizen science can empower individuals and communities, and demystify science as well as creating wonderful education opportunities.[15]

Whether this involves collecting data on birds in suburban backyards, analysing museum data or star charts, many environmental projects in particular have benefited from a volunteer army of citizen scientists getting involved. It also often reconnects people with nature and sparks imaginations and passions in the process of science.

And as a great validation of citizen science, in 2017 an exoplanet was discovered by a Darwin mechanic in a citizen scientist project that involved participants trawling through data from NASA's Kepler Space Telescope.[16] The project was a part of an ABC television event, Stargazing Live, that featured rock star British physicist Brian Cox.

In fact, Cox was so impressed with the find he said, 'In the seven years I've been making Stargazing Live this is the most significant scientific discovery we've ever made.'[16]

Citizen science can be done in a range of settings, from sitting at home on a computer to getting out with the whole family to count animals or plants in a particular area. All give people the opportunity to embrace their inner scientist, and that has to be intrinsically good – yes? – making a scientifically literate society.

You'd think so, and I have taken part in conversations with scientists who agree with that, but still have doubts about the quality of the data.

A study conducted by a team of seven researchers looked at over 1000 volunteers in citizen science, breaking them into six groups depending on their current engagement in science. They found that those with the highest levels of engagement were not only more likely to take part in citizen science projects – no real surprise there – but produced very good data. Comparing 1300 examples of citizen science-collected data with professional scientists' data, it was found first that while some researchers tend to be optimistic about the use of citizen science data in their qualitative discussions, only 50% to 60% of comparisons showed enough accuracy to meet the requirements for scientific research.

However, they also found that when there was more training of citizen scientists and when they had economic or health interests in the outcome of the research, they provided much higher quality data.[17]

So if you are getting involved in citizen science projects, make sure you do your training of participants well and try and ensure they have some personal buy-in.

Analysing several citizen science projects to find out what defined a successful project, a list from the Israel Institute of Technology includes:

- clear scientific goals
- clear educational goals
- working with an existing community
- user-friendly interfaces
- simple platforms for collecting data
- online videos, tutorials, and work guides
- social platform supports
- clear dissemination of results.[18]

An alternative list, from the University of New Hampshire, is:

- capitalise on connection to place
- identify and meet potential volunteers
- fulfill the volunteers' needs
- keep protocols simple
- make it fun
- people enjoy people
- provide periodic opportunities to renew (or not)
- record your success
- thank and reward participants
- share your results
- evaluate your project.[19]

So what is poor engagement?

There is always a flip-side to every coin, and a study of engagement options in regional Victoria in Australia, dealing with flooding, looked at barriers to effective engagement from poorly planned or executed engagement strategies, and at the result of efforts to fix the problems – shown in Table 14.4.

There are many reasons why potentially good engagements become poor engagements, and many of them might be out of your control, but the three cardinal sins of failed engagements are:

- making it more about marketing or trying to convert a stakeholder group to another's way of thinking
- developing the engagement in isolation from all the stakeholders needing to be engaged with
- making no impact on policy or technology development.

Table 14.4. Barriers to engagement and strategies to overcome them[20]

Problem	Strategies	Outcomes
Poor community engagement process through unskilled consultants. Stakeholders felt that level of community consultation was inadequate and not desirable.	Consultants held an additional meeting for a preview of draft strategy.	Stakeholders reported that they wouldn't expect any changes to be made at this late stage.
Key stakeholders were felt to have been excluded and important local knowledge ignored.	Consultants demonstrated some flexibility in responding to concerns.	Stakeholders were sceptical that their input would be valued as reliable or authoritative.
Lack of information sharing between groups by not providing stakeholders with adequate information to contribute to discussion. Consultants had low expectations of stakeholders' interest.	No strategy recorded.	Stakeholders felt that their input was tokenistic.

This last one is very important and is where many theoretical and research models fall down. They develop a great model and everyone signs up and they have a great engagement, but the powers that be in government or wherever else shrug at the findings and it makes no great impact.

So let's go back to Table 14.1, at the start of this chapter, and see how it should look, surviving Google translator's bullshit filter (see Table 14.5).

Table 14.5. Translating engagement objectives to what was actually meant

Original	Translated to what was actually meant
'We need more community engagement with our target audiences.'	'We need to develop some engagement models to better interact with our audiences.'
'We want to undertake more public engagement'	'We want to really engage with the public and co-design what they most want us to deliver.'
'There is not sufficient community engagement on our programs.'	'We are not getting enough community participation in our engagement activities and we need to find out what will work better by asking the community.'

What to do with what you now know

Engagement has become a vital part of science communication as it is now understood that for many issues, simply providing information is not enough. Communities want to be an active part of decisions that affect them.

There are many, many models of engagement that you could select from, but they are all based around giving the community you are working with some of the power – whether that is decision making power, or access to better information.

Five simple rules that will get you over the line in most instances are:

1. Figure out what level of engagement you are going to do, and how much power the community you are engaging with is going to have in the process.
2. Be open up front about the level of engagement you are undertaking so there is no misalignment of your and the community's expectations.
3. When you meet with a community, validate any feelings of concern that may exist (or any other strong feelings).
4. Discuss people's differing values and frame the engagement around those values.
5. When people want scientific facts, they will ask for them. Don't feel you need to give them before then.

Key summary points

- There are many forms of community engagement in practice, ranging from informing to empowering, with greater levels of engagement as you move towards empowering.

- The underlying purpose of community engagement is to empower communities to make decisions. Any methods chosen to engage with them should support this.

- There are many models of engagement to adapt. Look for the one that will work best for you.

- Effective engagement can take a long time to achieve.

- Effective engagement can also take a long time to learn, and effective strategies should be pursued to provide adequate skills to those undertaking engagement strategies, such as becoming accredited with professional engagement associations.

15

P-values: Policy and politics

'In development research, to get a new discovery into policy and practice is just as important as the discovery itself.'

– Maureen O'Neil, former President and CEO of the International Development Research Centre

I remember sitting in an organisation management course one day and the tutor, who was putting everyone in the room to sleep, suddenly said something so profound that it stuck with me after everything else he said has long since faded away. He said that the prime objective of any organism was to preserve its life. That was a metaphor for organisational management. Your prime objective is to maintain your budget. Without that, nothing else matters because you will cease to exist.

And he was quite right. If you are a commercial body, the thing that sees your demise above anything else is lack of income. If you are a government agency, the same thing applies. If you are not funded, you cease to exist.

That is the reason many science communicators who work for public sector agencies find themselves increasingly undertaking marketing or stakeholder liaison tasks. In tight financial times it is considered a more important use of your skills to help raise funds. Think of it the way that your body draws all the blood into the core organs at times of crisis. Who needs fingers if you are freezing? You need that blood to keep your heart and your lungs functioning.

So communicate more effectively with policymakers and politicians and you stand a better chance of maintaining your funding – or not having it whittled away – and you can then get on with communicating science.

But there is a growing sense amongst both scientists and policymakers that science does not engage with policymakers well, and that scientific information is not really contributing what it should to science policy, particularly in controversial areas.[1] Yet one of the biggest challenges I hear scientists and science communicators say they face is how to make an impact on policymakers and politicians. Is the problem a lack of trying or a lack of trying the right strategies?

Whichever side you tend to fall on, policymakers and politicians are an important audience that you cannot afford to ignore. And yes, they can be quite difficult to reach, but they are important because they are generally the people who control your budgets – and therefore your future.

The line between science and politics

In spite of some scientists stating that their work should be apolitical, there have always been fairly blurred boundaries between science and politics, and scientists have a long history of acting as advisors to politicians, or directly advocating to governments.[2] An often-quoted example is Albert Einstein's letter to President Roosevelt in 1939, urging the US government to accelerate academic research on nuclear chain reactions. The letter ultimately led to the Manhattan Project's development of an atomic weapon.

Interestingly, six years later, Leo Szilard, who helped Einstein draft his letter to President Roosevelt, wrote to President Truman, expressing concerns about using the scientific work of the Manhattan Project for political purposes.[3]

But for many scientists, capturing the attention of politicians can prove a very difficult thing. In Australia, the Science Meets Parliament program run by the lobby group Science and Technology Australia, has proven effective in getting hundreds of scientists into the offices of politicians to explain their work and its importance. But given the number of issues a politician has to deal with and the short amount of attention possible to devote to each one, it is still not easy to really make the impact needed.

For – like it or not – scientists rarely have the political or lobbying clout of all the other men and women in business suits who are lined up to talk to politicians regularly. Most lobby groups are power tribes, and they think like policymakers and politicians, and they talk like them, and when they engage with them each side pretty-well knows what each other are talking about.

And let's be honest here, most scientists come from a completely different tribe that talks a different language and has different customs and beliefs. And as any tribe tends to preference the ideas of the people in their tribe, and believe they have the right way of thinking about things, it makes it harder for tribal outsiders.

We are as guilty of it as the policymaker power tribes. If we belong to the science fan boy and fan girl tribe (see Chapter 6), we surround ourselves with our tribe and engage in a social media war on the other tribes, shaking our virtual slings and arrows at them.

We might decry short-term thinking that places profits over sustainability, or real estate over research. Or maybe we can see the danger of species loss, or monocultures. Or we get some aggrieved smugness from knowing that we are the tribe who can see that sensible investment in research will lead to widespread social, environmental and individual benefits.

But there are a lot of other tribes who don't see the world the same way. Tribes who might not even think there is much of a pay-off from science research, and who think that science is neither interesting nor relevant. There are even those – heaven forbid – who feel science can't really explain all the things that are unexplainable in the world.

If you want confirmation of that, just have a closer look at the science-based policies of the majority of politicians on your local council, or in your state or federal governments. And you really do need to look at them – for they are power tribes.

Power tribes tend to engage mostly with other power tribes: industry bodies, financial advisors, large lobby groups, consumer groups, unions and so on. There are lots of such tribes out there – generally cashed up and well connected. But scientists are not one of them. True, we do have more power than some tribes, like perhaps the I'd Rather Be Fishing Tribe or the LOL Cats Tribe. But we're just not up there with the big power tribes.

So here's the issue put simply. How do we become a power tribe and actually influence things more?

Well there are really three key ways:

1. Infiltrate existing power tribes more effectively.
2. Grow the size of our tribe and make it more dominant and powerful numerically.
3. All of the above.

We need to find ways for other tribes to consider science thinking more often, and consider it a part of their *own* world view, by us framing it through *their* world view. We need to talk science without ever using the word science sometimes.

For as rock star ecologist and anthropologist Jared Diamond points out, there are more than enough examples over the centuries of power tribes who have ignored the evidence-based voices around them, and died out when their world views proved inconsistent with

When science effectively influences politics...

sustainability.[4] Just as there are examples of tribes of evidence, who could see the problems facing them but weren't in a position of power to do anything about it (sound familiar?).

So who are policymakers exactly?

I've had the good fortune to work in several government agencies and work closely alongside policymakers and I can report that in fact they are pretty similar to you and me in most aspects. Some of them even have science degrees. But of course, many have education and experiences that are outside of science – and this, with the nature of their job, shapes their outlook on their professional life. And shapes their outlook on your professional life!

But more importantly, the nature of their job shapes their outlook on life.

Their job is a constant flow of many, many inputs that they need to consider, including economic, political, electoral, budgetary and media pressures, government preferences, lobby group preferences, key influencer opinions, industry perspectives, other policy directions, election commitments – and a need to have outputs that work for the political world they operate in.

While policymakers don't always have the scientific expertise needed, they generally have the policy expertise needed. And while you might hear the expression 'evidence-based policy' a lot, the truth is it is really means 'evidence-influenced policy'.

Sir Peter Gluckman, the outspoken former Chief Scientist of New Zealand, says policy is all about trade-offs, and those trade-offs matter. There are often political choices between long-term benefits and short-term electoral risks. However, he also says that in a post-expert, post-truth, post-trust world, making difficult decisions is harder and the policy challenge has become how to achieve consensus amongst divergent world views.

This can be particularly difficult for science controversies as there are not just contested and divergent world views, but there may be well organised groups like corporations, advocacy groups and NGOs who have their own agendas and interest in getting a different policy outcome to you. And getting their desired policy outcome may well be more important to them than any accurate understanding of the science.[5]

This forces some scientists to become lobbyists or advocates for a position themselves. For issues like infant vaccination this is easier to justify, but if it is about where money should be invested, it is a little harder. There are many political decisions that scientists would like to influence decisions on, but if they are inherently political, then scientists alone cannot answer them.[6]

What is the best way to reach policymakers?

The simple answer to that is – make yourself relevant to their needs. Fortunately, many scientists and science communicators have learned the hard way how to best reach policymakers and have distilled their experience into useful tips. Here are a few of them.

According to Nick Hillman, who is director of the UK Higher Education Policy Institute, and a former political adviser, the top 10 things to know in communicating with policymakers or politicians are:

1. Recognise the competing claims made on politicians' time.
2. Maintain your intellectual humility, as there is often contested evidence and policymakers don't wish to be swayed by emotional arguments.
3. Strike while your work is topical, not when it's completed and published.
4. Don't assume your published work will be read.
5. Watch the merry-go-round of staff moving from position to position, removing consistency of contacts, and wince.
6. If you don't like the weather, wait a while. Be patient. Things change.
7. Write for the ignorant but intelligent reader.
8. Be constructive. Too many policy papers have strong data, but a weak or insufficient conclusion.
9. Make your mark on the big picture. Policy recommendations that fit with the government's current narrative are more likely to be accepted.
10. Remember that no one remembers.[7]

According to Paul Cairney and Richard Kwiatkowski, also from the UK, insights from psychology and policy studies can be used to help communicate effectively with policymakers or politicians. They recommend three key things:

1. **Be concise.** Make your evidence concise to minimise its cognitive burden, and frame your conclusions rather than expecting your evidence to speak for itself.
2. **Identify the right time to exploit 'windows of opportunity'.** There is often a right time to influence a person, when the political conditions are just right.
3. **Engage with real world policymaking rather than waiting for a 'rational' and orderly process to appear.** To 'speak truth to power' without establishing trust can be very counterproductive.[8]

Former Australian Chief Scientist, Ian Chubb has stated that to succeed in influencing politicians you need to rely on 3 Ps (more P-factors!):

1. passion
2. persistence
3. patience.

What is a policy brief?

A policy brief is the type of document that policy people generally feel comfortable reading. The same way scientists feel comfortable reading journal papers. If you've never seen a policy brief it is worth finding one and seeing how they are written for the government or government agency you wish to communicate with.

You will find a policy brief is a concise, standalone document focusing on a particular issue requiring policy attention that:

*   explains and conveys the 'urgency' of the issue
*   presents policy recommendations or implications on the issue
*   gives evidence to support the reasoning behind those recommendations
*   points the reader to additional resources on the issue.
    Policy briefs should also be:
*   concise
*   accessible
*   practical
*   professional, but not academic
*   based on evidence
*   visual appealing (using graphs and charts as needed).[10]

But don't worry if you don't manage all of these. Very few policy briefs that are written by professional policymakers manage to get them all in a policy brief either.

Science meets Parliament

In Australia there is a very effective engagement model whereby roughly 200 scientists descend on Parliament House each year in the nation's capital, Canberra – not to protest, but to meet with politicians and learn about how politics works. It is similar to the US model of 'Hill Days', where scientists descend on Capitol Hill.

The program, which has been run by Science and Technology Australia for almost two decades, takes place over two days, and provides an opportunity for building understanding and connections between federal parliamentarians and those working in science and technology. It also seeks to ensure that science stays on the political agenda.[11]

Scientists who participate receive skills training in things like how policy works, and how to pitch your science in 45 seconds – something that aligns with the political need to get to the guts of an issue quickly.

A participant in the 2018 program, Hannah Brown, who conducts research into early pregnancy health, said the most valuable lesson she learned from attending was to remember that even though politicians are passionate about their job and their responsibilities, they are humans with complex lives and other interests too.[12]

He has said it is a long game and one that you should not expect to get significant impacts in easily nor quickly.[9]

The role of community engagement in policy formulation

We live in an era where most policy debates relevant to science or technologies are not simply science or technical issues, and need to be decided at the intersection of politics, values and expert knowledge. This needs better engagement methods to allow for discussions rather than information battles.

Added to this, the internet has enabled more members of the public to become more informed on issues of interest to them – indeed, they can often be better informed than policymakers or politicians. This is a fact of modern life that politicians disregard at

The five Ss that inhibit policymaking

British politician Vincent Cable has said that policymakers are practically incapable of using research-based evidence because they are limited by the five Ss:

1. **Speed:** they have to make decisions fast.
2. **Superficiality:** they cover a wide brief.
3. **Spin:** they have to stick to a decision, at least for a reasonable period of time.
4. **Secrecy:** many policy discussions have to be held in secret.
5. **Scientific ignorance:** few policymakers are scientists, and don't understand the scientific concept of testing a hypothesis.[13]

Case study: Why science is rarely taken into consideration in political decisions

Sorry scientists, but here's a hard truth – when it comes to most policy decisions, science is rarely a major consideration.

As an example, the state of New South Wales in Australia recently decided not to cull wild horses in the high country national parks. The evidence was clear that wild horses are an invasive species causing great damage to fragile alpine areas, and this damage further threatened endangered species, such as the much-loved tiny corroboree frog. Many scientific reports have been released demonstrating this, arguing there needs to be a cull of the horses to reduce their numbers and reduce their negative impacts.

But up against the scientific evidence we have the emotional appeal of wild horses. They represent that slice of Australian heritage embodied in the era of frontiersmen and women, and nationalistic poetry and films.

It is also useful to know that the State Parliament Bill to prevent wild horses being culled was introduced into Parliament by the Deputy Premier of New South Wales, John Barilaro, whose electorate of Monaro covers the highland national parks. Home of those who consider wild horses a part of their heritage, and descendants of the settler families who established cattle runs in the high country – largely on horseback.

Groups like the Australian Brumby Alliance have been outspoken about the importance of wild horses to our national heritage, with statements like:

It's magic. It's just a wonderful feeling; you just feel amazed at this majestic horse that can keep itself going in the park without any human interference. It uplifts my spirit.[14]

By comparison, scientists' rhetoric tends to be like:

Horses have been present in the Snowy Mountains since the 1830s when Europeans first explored the region. Substantial transhumance grazing (i.e. the annual movement of stock and stockmen to summer pastures in the High Country) of cattle and sheep soon followed and continued for more than 150 years.[15]

And as we should know, in politics decisions are informed by a large number of different factors, including: economic factors, interest group lobbying, political ideology, media stories – and scientific evidence.

Of course, most scientists would see the weight of scientific input as being stronger than all the others – or maybe even at least of equal weighting. But that's not the world we live in, is it?

If we had a spectrum of the emotive and electoral sensitivity of different inputs to policy, almost everything listed above would lie on the side of having high sensitivity – except for scientific evidence, which would be all alone on the other side having low electoral and emotional impact (see Fig. 15.1).

That means for issues that are not emotional or electorally sensitive, there's a good chance that the science input will count for something. But if the issue is being dominated by emotions and is electorally sensitive in any way – sorry science.

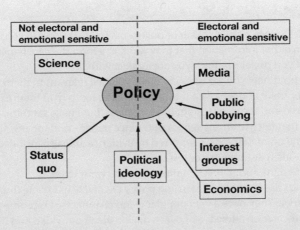

Fig. 15.1. Inputs to policy

I mean, which narrative do you think is the most powerful?

Evidence shows that wild horses are damaging sensitive environments and they need to be culled, preferably being shot from helicopters.

Versus

Wild horses are an iconic part of Australian heritage that reflect the Australian spirit, and it is cruel and inhumane to slaughter them.

The science narrative above actually plays into an emotional response against it, as we know that there are strong preferences for non-lethal control methods of larger invasive animals – especially among the urban public, who live a long, long way from where such animals roam through sensitive environments.

The same type of thing plays out in many contentious issues, where there is a conflict between scientific evidence versus emotional responses – whether the topic is climate change, coal seam gas mining, vaccination or embryonic stem cells (yeah, yeah, the usual suspects), emotions far outweigh the scientific evidence.

What is a scientist or a science communicator to do, given that they are often unable to play the emotional game to counter emotive arguments? Are you going to be perpetually out-played when trying to make some impact on policy?

Well not necessarily. It is possible to reframe your arguments to incorporate some element of the opposing emotive arguments. In this it might be possible to frame messages around putting the welfare of the wild horses first. Pushing to keep their numbers down so that wild horses won't grow to the point that it threatens their own wellbeing.

Frame messages that make you a wild horse lover, not a wild horse hater.

And above all, if you are embroiled in a policy debate, don't rely just on the evidence alone. Emotion and electoral sensitivity will usually be more important, and you will need to find some way to address them too.

What to do with what you now know

So you want to influence a policymaker or politician? The first thing to consider is that you need to give him or her information that they want, in a form they are familiar with – not information that you prefer in a form you are familiar with.

Then think about how you are going to make yourself relevant to them. What is their operating environment and what are they most looking for to help them in their job? Some things you can ask, and some you might have to guess, but I have sat through too many policy briefings that felt like an academic conference paper – and too few that felt like they knew to get to the point of things, supported by evidence and also included emotion and political capital – all in five minutes or less.

Have a careful think about which of those two you want to be.

An interesting case study on influencing policy creatively has come from the Australian Academy of Science. They have a very popular video channel that is promoted primarily on Facebook. They receive an audience of over a million views per video and they can make a short film on the impacts of climate change and take it to politicians with the two-fold objective of the politicians being engaged with the visual data and the knowledge that this message was actively watched by over a million people giving it more clout.

their peril and often accounts for the reason political decision makers so often get it wrong when trying to push a particular line, or spin a story related to a policy decision.[16]

Therefore, sophisticated public outreach and engagement are essential to overcoming perceptual gridlock on issues like climate change, or for involving the public in societal decisions about plant biotechnology and nanotechnology, or for effectively engaging with stakeholders and a wider public on almost any issue really.[17]

The point is that science communication can make itself much more valuable to policymaking by providing the type of public engagements that can come up with policy recommendations that align with the public's expectations or preferences, and avoid costly political turmoil.

Key summary points

- Policy decisions are often about trade-offs.

- Policymakers often seek to make evidence-based decisions, but because they have to manage so many different inputs from so many different areas, they are more likely to be make 'evidence-influenced' decisions.

- Knowing what policymakers are seeking, and providing it to them in a form they are familiar with, helps you reach them.

- There is a role for science communicators to undertake public engagement activities to help the development of policy that aligns better with public needs and wishes.

16

Evaluation: Metrics, damn metrics and statistics

'The single biggest problem in communication is the illusion that it has occurred.'

– George Bernard Shaw, author

I know we've talked a little about evaluating your science communication activities already – but now we're going to get a lot more serious about it.

Watching the way some science communication activities are evaluated makes me think of the story of the man who was standing around under the street light searching all over the ground. Another guy comes along and asks, 'What are you looking for?'

'I dropped my car keys,' the first guy says.

'I'll help you look for them,' the second guy says.

They search around for about 10 minutes and the second guy says, 'I can't see them anywhere. Are you sure you dropped them here?'

'Oh no,' the first guy says, 'I dropped them out there in the dark somewhere.'

'Then why on earth are you searching over here?'

'Because there's much more light over here!'

Even if you don't find the story funny (what's wrong with you?), you'll get the point of it – too many science communication activities are measuring things under the street light rather than out in the dark where things are harder to measure. Like measuring bums on seats rather than measuring if any understanding of the science has occurred. Like measuring how many people take a brochure rather than measuring how many actually read it. Like measuring how many people go 'Wow!' when you blow shit up rather than understanding the principles of chemistry or physics behind the explosion.

Which leads me to ask a big question: why are evaluations so important to scientific research and yet are so often forgotten or tacked on as an afterthought to science communication?

What the researchers say

But is it really that big an issue? Not if you check it out in most books on science communication – because they won't even have a section on evaluation!

Let's see what some of the experts and researchers say.

First, according to Eric Jensen at the University of Warwick in England, even in the best-resourced science communication institutions, poor quality evaluation methods are routinely employed. He said, 'This leads to questionable data, specious conclusions and stunted growth in the quality and effectiveness of science communication practice.'[1]

Ouch!

Or a US paper looking at the evidence-based approaches to interactive health communication found communicators often forgo evaluation as an unnecessary and costly step that merely demonstrates that what they believe that they already know.[2]

And a UK review of many museums found, 'evidence used to suggest learning or particular forms of learning can appear fragile at best'.[3]

Going back over a decade earlier, another UK study that looked at a review of the literature on evaluations of public awareness of science initiatives, found that many of them were not formally written up and fewer were evaluated against their aims. [4] You would think that might trigger some progress.

But evidently not. Eric Jensen once more, because he writes so much and so well on this, says that in lieu of robust evaluation techniques, organisations conduct anecdote gathering exercises, focused on eliciting positive accounts of how wonderful their program is.[5]

The question I put, over and over, to science communicators who are whooping up their latest event and celebrating how many people turned out to see it is, 'Yes, but how do you know you are actually making an impact?'

It is all too easy to believe you are when all those who took part in it feel so good about it. You can easily get carried away by the cheers of the audience who have gotten carried away when you exploded something.

But, as you wipe up the mess off the floor, ask yourself if you actually contributed to science literacy, and if so, how can you measure that impact? Particularly the long-term impact. If we don't measure that, we risk becoming our own worst enemies and going with what we think is going to work because we like it.

So what might a good evaluation look like?

In an effort to come up with a standard evaluation for the diversity of science communication activities that exist, the Australian Government science communication initiative, Inspiring Australia, developed a guide for evaluation.[6] Unfortunately I have seen people look at it and find it too complicated for their needs, which is the problem in trying to find one solution that works for everyone.

Marina Joubert from Stellenbosch University in South Africa, and a highly-regarded science communicator, has produced a practical guide of key principles for evaluating science communication projects (see box on page 144).

She says above all, the trick to evaluating a science communication project is to plan carefully. Also, learn from your mistakes.[7]

She acknowledges that communicating science to the public can encompass a very diverse array of approaches, including public talks, debates, exhibitions, publications, science theatre, television documentaries and citizen projects like consensus conferences. And audiences who attend these different events can range from young children and teenagers to parents, business or political leaders.

Finding a single evaluation tool for such a diversity of people attending such a diversity of events can be difficult.

But it can be useful to go back to the overall driving objective of the event (see Chapter 4). Is it being undertaken to encourage young people to consider science careers? Or maybe it is being done to support dialogue and informed decision making about science and society issues? Or perhaps to promote public acceptance of new technologies, hoping to change attitudes or behaviours? Or maybe it is just being done to foster a science 'culture'.

If so, these are the things that you really should be measuring. Not just how many people attended the event – but how it has contributed to your overall goal.

What is a goal and what is an objective?

As described in Chapter 4 a goal is the big picture thing you are trying to achieve – like increasing science literacy or encouraging more people into science careers. An objective is much more specific. It is also measurable. Such as getting 100 people to attend an event, or raising awareness of your activities by 10%.

South African science communicator Marina Joubert says:

> Don't shy away from evaluation because it seems expensive or difficult, or because you fear being seen in a negative light. And don't leave evaluation until after the project is over. Unless you plan for evaluation from the start, you won't have the people, instruments or resources in place to do it. You could miss critical opportunities to gather data and could end up with insufficient evidence about your project's impact.[7]

Not even trying	Just guessing	Measuring smiles	Asking colleagues	My mum loved it	Bloody brilliant	Top marks

Evaluations are improved by having the right measurement tools

She also says it can be very useful to involve your co-workers in designing the evaluation plan, so that they feel shared ownership, and thus, responsibility for it, helping to ensure its success. In other words, don't just work top down. Get everyone involved whether they are scientist, science communicator or manager.

Non-traditional evaluation methods

Because it is often too easy to say it is too hard, there are many alternative metrics (alt metrics) that you can employ. They sometimes involve a bit of lateral thinking but can be used to capture data in different ways. These include:

- **Snapshot interviews:** which are short face-to-face interviews rarely taking more than two minutes, but many of them can be performed in a short time.

Key evaluation principles

- Quantitative evaluation is generally measuring numbers, like the number of people who attend an event, or the number of visits to a website.
- Qualitative evaluation uses interviews or questions and answers to build up more quality-based measures.
- The best evaluation happens before, during and after a project to test messages, measure how well something is happening, and then to measure its actual impact.
- Observational tools can include using photographs and video footage to record actual behaviours for analysis.
- Evaluating websites should include monitoring how people navigate your site and how much time they spend in particular places.
- Events are best evaluated with simple questionnaires. The simpler the better. Stick to your key questions only. For large events, use a tear the box form, rather than have people scrabbling for pens or pencils.
- Interviewing people immediately after an event gives you current impact information, but for longer-term impacts, get their details and contact them again some days later.
- Focus groups are small groups of people who are often specifically recruited to represent certain demographics, and to discuss an issue.
- Self-evaluation, if done honestly and critically, can complement external evaluations.
- Measuring if a project led to attitude or opinion change can require large-scale, expensive surveys, which are best done in collaboration with experienced social science researchers (though let's be honest, any evaluative technique that would benefit from an experienced social science researcher is worth getting input from one).
- Analysing media coverage can help indicate a project's wider impact.[7]

- **Graffiti walls:** where participants are invited to draw, write or stick notes onto a wall, or large piece of paper, giving their thoughts or feelings about an activity.
- **Feedback boards:** similar to graffiti walls, but with headings asking particular questions.[9]
- **Narratives:** collecting stories from people in their own words.[10]
- **Qualitative or mixed media feedback:** used by the performing arts use to measure success, can be adapted to science communication.[11]

Science centres and science events

Talking on evaluations, we need a special mention of science centres and science events. It should be no secret that many science centres find it really hard to effectively measure their impacts on visitors, yet all the while those who fund them are constantly seeking metrics. This has pushed many towards the lowest hanging fruit, and an emphasis on those things that they can measure (such as bums on seats) rather than the things that they should be measuring (such as impact).

It can be very tempting to fall into the trap of false causation. For instance, when looking at data that shows that the people who visit science centres regularly are then

Evaluation in practice

- Keep your evaluation simple, practical and user-friendly. Lengthy or complex survey forms may scare people away.
- Pre-test your evaluation tools to make sure they deliver the data that is useful for you and helps measure your objective.
- Make sure the language is appropriate for the audience.
- See if you can get some baseline data to begin with that has measured for instance, awareness of your program, so you can actually measure if there has been a change in awareness.
- Build on your findings by feeding your results into future activities.
- Share your results – including your mistakes – which can be valuable information for your team members, partners, collaborators and sponsors.[7]
- There are a growing number of new technologies that can be used for evaluations that include:
  ▶ visitor apps (iPhone and Android app)
  ▶ visitor insight dashboards with real-time results of automated analyses
  ▶ information kiosk touchscreens
  ▶ automated analysis of the content of social media data about the experience.[8]

more interested in science, it is perhaps worth considering that it is those that are more interested in science that visit science centres regularly.

Possibly the best evaluation of the impact of science centres is a 2014 report, *International Science Centre Impact Study*. It found that visiting a science centre

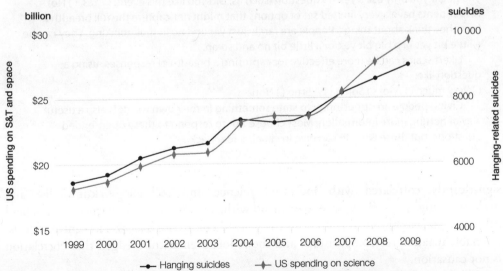

Fig. 16.1. Correlation is not causation: correlations between US spending on science, space and technology and suicides by hanging, strangulation and suffocation[13]

Communication evaluations to avoid

The types of evaluation that don't work, or favour biased answers, or allow the participant to tell you what they think you want to know, include:

- asking people whether they believe a communication would affect their behaviour
- relying on your instincts as to how well something worked, without testing that with evidence
- forgetting that anecdote + anecdote + anecdote ≠ evidence
- confusing qualitative and quantitative research and believing a focus group finding represents the wider population
- forgetting that you are probably not one of your target group and probably don't think like they do
- not learning from data you get as you evaluate something in action
- believing immediate post-event surveys correlate with long-term behaviour and attitude change
- relying on self-reported knowledge levels or intentions
- measuring social media – such as tweets – quantitatively, rather than their impact
- measuring column inches or centimetres of printed coverage, rather than measuring their impact.

Asking the right question

As in public attitude research, how you ask a question can determine what type of answer you get. If you only ask a yes/no question (such as: Did you like this event: ( ) Yes, ( ) No) – respondents have a very limited set of options that might not capture the full breadth of how they think about things. People are rarely just yes and no in their thinking. They can be a little bit yes or a big bit yes, or a little bit no and so on.

Likert scales can be more effective for capturing a breadth of responses, using a question like:

( ) Very much ( ) A lot ( ) Some ( ) A little ( ) None

Giving people an opportunity to state something in their own words is also a useful way of getting more information. Most choose not to respond to these open-ended questions, but those who do can give invaluable feedback.

significantly correlated with increased science and technology knowledge and understanding, as well as an engagement with, and interest in, science as a school subject. Data was obtained from both the general community, as well as centre visitors, to enable meaningful comparisons, and the report even acknowledged that correlation is not causation.[12]

While certainly not perfect, it is an honest attempt to measure something very difficult to measure.

What to do with what you now know

If you are serious about undertaking a good evaluation, plan it early in your communication and tie it to your objective – not just to what is the easiest thing to measure.

Too many evaluations, like too many objectives, are a bit wishy-washy and don't really prove much. Like measuring how many people came to an event rather than measuring how many people learned something new at the event. As much as you have a purpose for undertaking a science communication event or activity, those who take part in them have a purpose for doing so too.

A good evaluation finds out if their purpose was met, not just if yours was met.

There are many tools for collecting data, and if unsure, talk to someone who collects data regularly – they might surprise you with a very easy method that gives you just what you want. I attended a movie screening where everyone got a slip of paper with the program on it. Down the side was a rating scale and all the members of the audience had to do was put a tear against the number of their level of satisfaction and drop it in a special bin on their way out. The response was extremely high as the task was extremely easy.

And be prepared for findings you don't like – they are very useful for fine-tuning your activities.

Key summary points

- Evaluation is very important but is often approached as being in the 'too hard' basket for science communication activities, and too often ends up being tacked on the end rather than well planned.

- There are many good evaluation tools that exist, including apps and alternative evaluation metrics. Find them and use them. A good starting point would be those social scientists who use them regularly.

- When you have decided on an evaluation tool, test it to make sure it works for you.

- Don't fall into the trap of measuring the easy things to measure, like number of attendees, rather than actual impact – which you might measure by increase in knowledge though a simple survey.

WHEN THINGS GET HARD

17

I'm a believer! Understanding different beliefs

'I can believe things that are true and things that aren't true and I can believe things where nobody knows if they're true or not.'

– Neil Gaiman, *American Gods*

Beliefs are best thought of like those annoying little habits we all have – biting our nails, or doing some strange shrug, or a finger twitch thing – because they can be just as inexplicable as to how we picked them up, and just as difficult to get rid of.

For science communicators, understanding a person's beliefs, or belief systems, can be more important than understanding their attitudes – because attitudes can often be changed. Beliefs – very rarely.

So what do we mean by a belief?

It might be as simple as believing that people should not destroy their natural environment. Or that their sporting team is going to win their next game, regardless of their track record. Or it might be that UFOs have visited the earth and abducted people and done strange experiments on them.

The most important thing to know about beliefs though, is that people really believe them. Did I mention we are not talking about attitudes here? I *believe* I did.

And another thing about beliefs that we really need to know is that they are bullet-proof to being influenced by any facts that proves they are wrong. You can test that out by having a conversation in a bar with any avid sports fan, using statistics to tell them why their team is not going to win the next big game.

As has been mentioned earlier, the issue of climate change isn't about what you know. It's about who you are.[1]

Of course, they won't believe you. They will probably even put money down to bet against you. Because beliefs form a part of a person's world view and make up who they are. If you challenge someone's beliefs, you threaten not just their understanding of the world, but you threaten who they are.

Why else do climate change deniers, pro-gun control and anti-vaccine advocates act so crazy when they are shown data that proves their beliefs are wrong? That data is threatening their very person – and so it has to be dismissed as inaccurate, wrong or misleading.

And it is not too far an analogy to suggest that contradicting a person's general beliefs is similar to contradicting someone's religious beliefs. Certainly, it is useful if it helps you realise how sensitive people can be if they perceive their beliefs are being insulted, and how emotive their responses can be.

Understanding the breadth of beliefs across a community

There are a variety of different beliefs across any community. Some people, for instance, may believe that nature is sacred, or that multinational companies are not to be trusted, while others believe that economic progress is vital for humanity's success on the planet. Others might have more spiritual beliefs at the heart of their understandings of the world – such as the existence of angels, or spirit auras or astrology. **The breadth of beliefs across society can be very broad, and a scientific view of the world is really just another belief, competing with others.**

It is too easy for a scientist or science communicator to mock other's beliefs as inferior or wrong, as many members of sceptics organisations do, but that doesn't lead to paths to communicate. It blocks them. Better to understand what different beliefs exist, and take that as the reality of the environment you are working in.

Community values and beliefs should not be considered as something to be changed, rather as a path to understanding community perspectives.

There have been enough surveys that show that not only is there a significant gap between science-based views of the world and general community views, but that the science-based views are often a minority – quite removed from the general average community point of view. And the further the gap between them, the harder it is to understand different perspectives.

So while you might sit around over drinks with your friends or colleagues mocking those who believe in UFOs, there is a good chance they are sitting around over drinks mocking you too.

Why do such different beliefs exist?

We can blame our brains. We are wired to find meaning where meanings might not always exist. We are wired to not just search for, but to find, patterns that have meaning to us – like seeing a face on Mars if you look hard enough for it, or seeing Christ in the dark marks on a piece of toast or on a damp stain on the wall.

Let me give you an example.

On a recent trip to the Australian National Gallery in Canberra with my grown-up daughter and her boyfriend, we were detained by a guard in front of Jackson Pollock's *Blue Poles* painting. And you know that being detained by a guard anywhere is never going to be a good thing. But we weren't expecting this.

The painting, which is a fine example of abstract expressionism, was bought by the Australian Government in 1973 for $1.3 million (which in today's money is roughly a gigagazillion dollars), and it has been a media item for many years, with people either decreeing it as brilliant or an extravagant waste of money. Needless to say, everyone wants to come and see what all the fuss is about.

When it was exhibited at the Museum of Modern Art in New York in 1998–99, it was hung as the signature work of the exhibition, dominating the last gallery of the show, and was described as ending the show 'not with a whimper, but a bang'.[2]

The painting is a riot of dripped paint and bright colours, as Jackson Pollock is known for, with eight blue poles across the huge canvas. The colours leap between them like moving sparks of electricity. At least that is what I see when I sit there and look at it.

But this guard wanted to show us what he saw in the painting. There were, he told us, faces hidden in the painting. Then he painstakingly waved his hands around and used a pen on a brochure he had to show us the rough head of a man, a horse, and a pig – and even a koala.

And yeah, if we squinted our eyes and tilted our heads and looked really hard, it is conceivable that Jackson Pollock might have hidden the real vague rough approximations of the head of a man or even a horse in the painting. But a koala? Like he knew it was going to be purchased by Australia 17 years after he died?

But more to the point, I know there is a condition called facial pareidolia, whereby people see faces in patterns around them. My rational brain kicks in here and I had to tell the guard thanks very much for his interpretation as we wandered into the next gallery. We all rolled our eyes of course and laughed a bit – before remembering what Oscar Wilde wrote (probably not referring to museum guards with facial pareidolia, or other odd beliefs, but just as applicable to them). He said, 'Even bad poets are sincere.'

Well, that's what I had always *believed* he had said, but when I checked it, what he actually said was, 'All bad poetry springs from genuine feeling.'[3]

Which in itself is a good example of the way our brain reinterprets data like memories to better align with things that have meaning to us. So if I mock the guard for seeing faces in the abstract painting that he wanted to believe he could see, I have to mock myself too, for believing in the Oscar Wilde quote that I most wanted to believe in.

The sad fact is we are all guilty of it to some extent. Even me and even you.

The Mandela effect

There is actually a condition called 'the Mandela effect' – named after the South African freedom fighter Nelson Mandela, who died in prison in South Africa in the 1980s. Many people remember how they felt at the news and how moving his funeral ceremony on TV was. Except for the fact that he never died in prison in the 1980s.[4] He was released and went on to become the first black President of the country, not dying until 2013.

But many people swear they remember the news story of how he died in prison. The fact is that there are many misremembered or misconstrued facts that become engrained in our culture. Misquoted lines of poetry. Mistaken deaths and dates. Or mistaken scientific findings.

One of the biggest is the mistaken belief that there is scientific evidence that some childhood vaccines can cause autism. The original story was that in 1998 British doctor

Andrew Wakefield and several colleagues published a paper in the esteemed medical journal, *The Lancet*, that suggested that the measles, mumps and rubella (MMR) vaccine may 'predispose to behavioral regression and pervasive developmental disorder in children'.[5] Or autism.

Upon closer examination, however, a few more things emerged. First, the study was only on 12 children, hardly enough to be a useful sample size. Second, Wakefield failed to disclose that he had been funded by lawyers who had been engaged by parents in lawsuits against vaccine-producing companies. And third, it was determined that he had selectively chosen data to suit his argument.

As a result of this, his co-authors removed their names from the study, the paper was retracted, and Dr Wakefield was struck off the UK medical register for unethical behaviour and other misconduct.

But the story was out there and would not be struck off from people's minds. You can search around online and find just how many people still clearly believe in the study. And the fact that the 'establishment' have tried so hard to close the story down just proves there must be some truth in it. After all, big pharma and governments are putting all kinds of things in our food and water and trying to cover it up, right?

Well they are if you believe they are. And if you believe they are, you are going to find information to confirm that belief.

US journalist Alia E. Dastagir wrote an article for the esteemed US journal, *USA Today*, entitled, 'People trust science. So why don't they believe it?' – in which she said:

> Think about the way you search for information. If you're a new mom who believes vaccines cause autism – and a number of women in your mommy group do, too – are you searching for research that shows whether they actually do, or are you Googling 'vaccines cause autism' to find stories to affirm your belief?[6]

She also quoted political scientist Charles Taber, who said:

> You have a basic psychological tendency to perpetuate your own beliefs to really discount anything that runs against your own prior views.[6]

Added to that, the power of anecdote or story over fact is quite strong, and the basis for many confirmation biases is an anecdote. My aunt's neighbour's cousin's child got a vaccine and straight away started being autistic! The fact that is a dataset of only one, is discounted because of the emotional strength of the anecdote, confirming a particular belief.

So why do clever people believe stupid things?

What do you make of the fact that roughly every second person believes in psychic powers such as ESP? Or that one

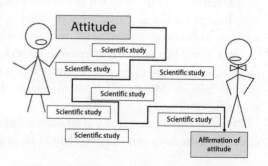

The determined path of confirmation bias

in three people believes in UFOs? Or that one in five believes in magic? Or that one in 10 say that astrology is 'very scientific', while three in 10 say it is 'sort of scientific', and only 50% say it is not at all scientific?[7]

Indeed, surveys conducted in Australia, the US and the UK indicate that about 80% of the population hold at least one paranormal belief.[8]

Or what are we to make of the faith being put in alternative medicines? A 2005 survey published in the *Medical Journal of Australia* stated that every second person is taking some form of alternative medicine – with one in four not even telling their doctor they are taking them.[9]

I'm sure we've all met somebody at a party trying to convince us of the benefits of the latest alternative therapies, which would be okay if it were harmless – but what happens when fringe medical beliefs lead to people dying from putting their trust in alternative therapies or faith healing, when traditional medicine could have saved them?

Steve Jobs is sometimes held up as someone whose cancer might have been treatable if he had chosen traditional surgery over alternative medicines much earlier.[10]

Coming back to vaccinations and autism, the US Centers for Disease Control found that one in five Americans believes that vaccines can cause autism, and two in five Americans have either delayed or refused vaccines for their child as a result of fears of their effects.[11]

And in Australia, according to the National Centre for Immunisation Research, before a national campaign linking vaccination with child payments, vaccination rates had been dropping to the point where only 85% of two-year-olds were covered – which is below the 90% rate needed to assure community-wide disease protection, and more likely to lead to outbreaks of fatal, but preventable, diseases.[12]

Non-scientific beliefs can be very strongly ingrained and are not likely to be influenced by any scientific facts.

Should we be concerned? Well only if we think that the dangers of non-science and pseudoscience are tangible and that widespread support for non-scientific beliefs can impede a society's ability to function, or compete, in an ever more complicated and science and technology-driven world.

The renowned Dr Ben Goldacre, UK author of *The Guardian* column and book, *Bad Science*, asks it as, 'Why do clever people believe stupid things?'

The answers, like many answers to complex problems, are – well – complex, but there have been enough scientific studies conducted to give us some insights into both what makes otherwise rational people have irrational beliefs, and what might be done to prevent it.

To better understand the factors that contribute to anti-science beliefs, and to help us know how they impact on science communication, we need to look at five key things:

1. scales of belief
2. mental shortcuts and motivated reasoning
3. the fear factor
4. the dangers of intuitive thinking
5. the uncertainty principle.

And we should also consider the values in society that allow anti-science messages to resonate with people. For just because you have a fringe idea and a website doesn't mean it will be picked up by people. It needs to have some resonance with existing beliefs to do this.

So the first on our list is:

1. Scales of belief

People don't divide into simple for and against camps on most things, despite them often being divided that way. One of my favourite sayings is: There are two types of people in the world – those who think you can divide everybody into two camps, and those who don't.

Scales of belief

There is usually a wide scale, or continuum (to use a good sciency word), of strengths of beliefs. So just because you believe in homeopathy and think that genetically engineered crops are unnatural, it doesn't mean that you don't prescribe to a scientific view of the world on other things. But the further along the continuum you travel towards the extreme anti-science thinking end, the more science-thinking is rejected – and people at that end are very unlikely to ever shift their position.

So first we need to understand how far along the scale of belief a person is, for the closer to either end a person is, that's where the rust forms around their feet and the less likely they are ever going to change their attitudes or beliefs.

This brings us the second point:

2. Mental shortcuts and motivated reasoning

In plain English this means the human tendency to take mental shortcuts. Social scientists call these heuristics (another good sciency word). It's the way we respond to rapid and complex information being fired at us. We need to quickly sort it into categories, and an easy way to do this is to sort it according to our existing belief systems or values.

Nobel Prize winner and best-selling author, Daniel Kahneman, captures the ideas well in his book *Thinking, Fast and Slow*. He describes slow thinking as being analytical but using a lot of brain energy, and fast thinking as being based on our mental shortcuts and strongly-favoured for using less of our precious brain energy. But it is more prone to errors. A lot of people have read his book and gotten that message. But another key message, less often cited, is that while we can often spot errors in thinking in others – we can almost never spot them in ourselves.[13]

This holds true for beliefs about GM foods, nanotechnology safety, climate change, your favourite political party, etc. – and the more complex the issue, the more likely people will make decisions based on beliefs or values and make errors. But you would never do that yourself, would you!

Would you?

Dan Kahan of Yale has studied what he calls the cultural cognition effect (great sciencey words), which put simply, argues that our values are more strongly going to influence our attitudes than any standard demographic like age, gender, race or socio-political status.[14]

The human brain has a preference to conserve energy

An interesting point here is to examine the differences in attitudes between the political left and right and their values and attitudes towards GM crops and climate change. Traditionally the left endorses the science behind climate change but is dismissive of the science behind GM crops, while the right tends to be supportive of the science behind GM crops and dismissive of that of climate change. Why? One accords with a world view of continued industrialisation towards progress and one accords with a world view about a need to value nature. If you look around enough you will probably find there is some topic that we are each going to become a science denier on as it disagrees with our strongly-held values.

In an ideal world we would look at different information and analyse it carefully and make up our mind. But that doesn't work when we don't have the motive or ability to do this. We are increasingly time poor in an increasingly data rich world, that forces us to make mental shortcuts more often, drawing upon

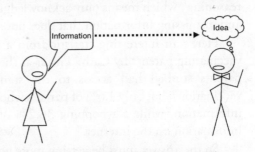

The imagined path of information to attitude

whatever existing knowledge we have (all too often from the media rather than from formal education), or falling back on our basic beliefs.

And in the age of the infonet (yeah, I just made that one up) the information-communication flows are entirely different to what we may have been used to, even a decade ago.

We all know that the promise of the internet to provide us with a wealth of information to make us smarter and better was akin to the early hopes that television would make us more educated and could teach us many languages, etc. ... Instead we are better at watching people dance and sing and hook up or cook via TV, and online we are swamped in a tsunami of irrelevant data of cats and game walk-throughs and meals and algorithms that actually filter to

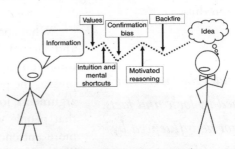

The actual path of information to attitude

selectively give us information that supports our existing beliefs as determined through our search history.

The internet is not fully to blame – it is just a channel for information – but the sheer amount of data of dubious credibility that is available from the internet, that doesn't readily distinguish between comment and research and blog and news, has changed the relationship between information and knowledge radically.

Where once we might have started with the germ of a wacky idea and sought to check its validity with experts such as teachers, or even by reading an encyclopedia, we now have the ability to pretty well find a community of people somewhere in the world with similar wacky ideas, never tested by an expert.

And through ongoing affirmation and reinforcement of wacky ideas, they can become beliefs, and then they don't easily move. If you doubt this, just Google: 'The Royal Family are Seven Foot Shape-Shifting Aliens' – and look at the sheer amount of confirmation on different sites about this (see page 160).

Access to the enormous breadth of opinions on the internet has revealed that people, when swamped with information and using mental shortcuts, turn to 'motivated reasoning', which means only acknowledging information that accords with our beliefs, and dismissing information that does not accord with them.

Here's an interesting statistic from a 2003 PhD study into vaccinating and non-vaccinating parents by Cathy Frazer at the Australian National University. All of those parents studied had access to the standard Health Department publications on vaccinations, but only 1.6% of parents who chose to vaccinate used the internet for more information, while a whopping 36.2% of non-immunising parents sought out other information on the internet.[15]

So the answer must be getting more good facts out there, right?

Maybe not.

Brendan Nyhan at the University of Michigan undertook a study that found that when people were shown information that proved that their beliefs were wrong, they actually become more entrenched in their original beliefs. This is known in the business as 'backfire'.[16]

And what's more, highly intelligent people tend to suffer backfire more than less intelligent people do – which is bit a like a vaccine, making us immune to any facts that our counter to our strongly held beliefs (though Chapter 24 provides an alternate take on this).

Attitudes that were not formed by logic and facts, cannot be influenced by logic and facts.

What about just providing the public with more balanced and factual information so they can make up their own minds on things? Well that can be a problem too, when you present the public with both sides of a story, giving them the arguments for and the arguments against. Research shows that people with an existing attitude tend to just become more entrenched in their original viewpoint, and are less likely to see the merit of other viewpoints.

This research, conducted by Dr Andrew Binder, found that most people, when faced with an issue related to science and technology, fairly quickly adopted an initial position of support or opposition, based on a variety of mental shortcuts and predisposed beliefs. And then, the more people with opposing points of view talked about a divisive science and technology issue, the less likely people from different camps would agree on any issue or even see it the same way.

Dr Binder stated:

> This is problematic because it suggests that individuals are very selective in choosing their discussion partners and hearing only what they want to hear during discussions of controversial issues.[17]

Now, as the media tend to always look for balance in their stories, particularly contentious stories, giving one side of the argument on say climate change or GM foods and then giving the other side can actually exacerbate this problem of polarised extreme opinions.

And an added thing here to note is that the more emotive somebody is about a topic, the less likely they are to be influenced by any facts and figures. So look around and see who is playing the 'scare card' and whipping up a bit of emotional concern about things. The more agitated, scared, upset or angry people are, the more receptive to emotive messages they are and less receptive to facts.

Which brings us to our third point:

3. The fear factor!

Franklin D Roosevelt said, 'We have nothing to fear but fear itself.'

If only.

When it comes to the fear factor, even the physical layout of our brains is working against us. Studies have shown that sensory inputs are sorted by the thalamus, sort of the brain's post office, and sent on to different parts of the brain to respond to. But the amygdala, which is the 'Danger, danger Will Robinson!' part of the brain is located right next door, and it is believed to get messages more quickly than other parts of the brain that are further away or lack as many neural connections. Our higher order thinking and decision making happens in the prefrontal cortex, which is not so closely connected.[18]

As researcher David Ropeik says:

> Both the physical architecture and biochemistry of the brain insure that emotion and instinct have the upper hand over reason and rationality. Before you know you are afraid, you are. The inescapable truth is that, when it comes to risk, we are hardwired to feel first and think second.[18]

Our fourth topic to know is:

4. The dangers of intuitive thinking

Jenny McCarthy, celebrity leader of the anti-vaccination movement in the US, says she bases vaccine rejectionism on her intuition. Likewise, many alternative therapists

Case study: Australia does not really exist

I often tell members of sceptics groups not to mock nor denigrate those with alternative beliefs, as it just isolates them. But there are a few groups I might make an exception for. One is the 'Flat Earthers' who, according to a recent story, believe that Australia doesn't exist and the people who claim to live there are actors paid by NASA.[19]

The story got quite a bit of coverage when the idea was put forward at a meeting of the Flat Earthers in Birmingham, in the UK, where 'over 200 people came together to confirm to each other that the Earth is nothing more than a giant pancake'.[19]

You can actually trace the story of Fake Australia back to a post started on Reddit in 2017, by one Shelley Floryd, which has since gained a lot of support. A Flat Earther Facebook post, for instance, states that not only is Australia a hoax, but that in the 18th and 19th centuries Britain did not move any convicts there, rather they were loaded off the ships into the waters, drowning before they could see land ever again.

'It's a cover-up for one of the greatest mass murders in history.'

And what's more, any Australians you might see in pictures or on TV are in reality computer-generated characters (sorry Hugh Jackman), and if you've ever been Down Under yourself: 'you're terribly wrong'. Because airlines have taken you to parts of South America instead, as the airlines are in on the plot!

Other beliefs put forward by the Flat Earthers include:

- gravity doesn't exist and the only true force in nature is electromagnetism
- NASA images of the earth from space are fakes

But ask yourself ...

But here's the thing – while you might be having a bit of a laugh and thinking that Flat Earthers must be really dumb, you need to ask yourself if you really wanted to believe this story, just because it made the Flat Earthers seem so dumb? Because a fact check of the story – not just finding sites that repost the story to seemingly confirm it – reveals that the original Facebook post was a joke.[20] But a joke that a lot of people really wanted to believe and so they fell for it.

But that's how beliefs work, right?

advocate that people should put their trust in their own intuition to justify their choices. So if people chose not to vaccinate their children, it's not because they are stupid – they don't vaccinate because their fear of vaccinating has become stronger than their fear of not vaccinating.

The diseases that we vaccinate against are unknown and unseen. We no longer see children dying from whooping cough or suffering from polio – but what we do see are strongly promoted stories of children suffering autism and other conditions as a result of vaccinations. So no matter how small a risk they are, they are visible and known – and therefore much more prominent a risk.

An outbreak of whooping cough or measles might change all this, of course, which is a dangerous possibility in some parts of the developed world at the moment. Health officials in the US at the turn of the 20th century talked about the need for a 'fool killer'

– which was an outbreak of smallpox devastating enough to convince people of the need for vaccinations.[11]

But in the absence of that, many of us are relying on our intuition, which tells us that autism or other side effects are a possible danger, and the risk of the disease being vaccinated for is a lesser danger – even though evidence from research shows the exact opposite.

Marcia McNutt, a former editor of the journal *Science*, and President of the National Academy of Sciences, said, 'Being a scientist only means that when I have an intuition about something, I test that intuition, and see if I'm right'.[6]

Our intuition has served us well for tens of thousands of years, stopping us from stepping out of the safe cave into the dangerous dark of night and so on, but it can be very unsuited for the 21st century where we no longer live in caves, and it can lead to false beliefs such as superstitions, paranormal phenomena and pseudoscience.

Which brings us to the fifth thing we need to know:

5. Uncertainty and control

At the heart of a lot of our non-science beliefs is control. We live in an ever uncertain and more out-of-control world. But superstitious beliefs and pseudoscience can give people a sense of control and certainty – through simple answers – which reduces our levels of stress at not having control. This is a necessary adaptive mechanism and something we tend to be wired to seek out. But here's the thing: science is predominantly based on uncertainty, while fringe beliefs are often based on providing more certainty. And as such, certainty wins out!

So what are we to do?

Here's the issue boiled down simply: many of us are living in a technology-driven world that our innate instinctive reasoning poorly equips us for. And there is no simple solution to address that.

But good science education can help.

There is some evidence that adults with more science training will more often reject astrology or lucky numbers, and more often accept evolution.[22] Likewise a 2002 PhD study by Alyssa Taylor in Canada found that a course on critical thinking led to a significant decline in belief in paranormal claims.[23]

However, we need to temper this finding with the results of a Canadian study that found that a 13-week lecture course critically examining paranormal beliefs led to a reduction in belief from 56% to 41% – but that figure crept back up to 50% a year later.[23]

So we need to educate people before attitudes and beliefs are strongly formed. And in this it is more important to teach them **how** to think than **what** to think. One way to make people a little more bullet-proof to pseudoscience is to effectively teach the values and ways of science thinking while still young – before alternative belief systems have formed.

There is no guarantee it will work with everybody, but without attempting to address it we leave too many people at the mercy of our mental shortcuts, our fears, our

David Icke and the 12-foot alien illuminati

We all need beliefs to help us make sense of the world, even if those beliefs are sometimes a bit extreme.

Author Jon Ronson tells the story of David Icke, a one-time British goalkeeper and TV sports show host, who became a self-proclaimed son of God, and now tours the globe explaining how the British royal family and key American politicians are actually shape-shifting aliens.

Check him out online if you want a wild ride down the online rabbit hole.

Ronson tells that on one occasion David Icke was planning to travel to Canada to give lectures, and a groundswell of protest started growing against his visit. Not because he believed in shape-shifting reptilian aliens, but because his rhetoric was clearly a metaphor for Jews.

He tells the story of one committee who had assembled to protest the visit. 'David Icke represents a political threat,' a speaker at the meeting said, 'His writings are anti-Semitic. David Icke states that the global elite, the illuminati who dominate every aspect of our lives, are genetically descended from an extraterrestrial race of reptiles who came to earth some time ago in the form of humans, who are capable of changing their shape, who engage in ritual child sacrifice, who drink blood.'[21]

But they weren't fooled by the metaphors.

Ron Jonsson however, who has interviewed David Icke many times, was not quite so sure. He wrote: 'I guessed that when he said that 12-foot lizards secretly ruled the world, he really was referring to lizards.'

Indeed, David Icke has declared that the well-known forgery, the Protocols of Zion, is evidence not of a Jewish plot, but of a reptilian plot of illuminati lizards.

So when David Icke arrived in Canada, officials ushered him into an interview room. They wanted to know more about his anti-Semitic stance before approving his entry into the country.

David Icke then told them of his theories that there was a genetic network that operates through all races, and that there was a certain elite bloodline that was a fusion of human and reptilian genes.

The immigration officers, after four hours of questioning, also concluded that when David Icke said lizards, he meant lizards.

Protests continued to follow his tour of Canada, as he did radio and TV spots and bookshop visits – not because it was too outrageous that anyone might believe that many of the world's leaders were descended from shape-shifting reptiles though. Because for many, that it was all a metaphor for Jews was the only story that fitted their own belief systems.

intuitions and our desire for simple answers to complex issues. This will not serve us very well for the challenges of the future – particularly when marketing gurus such as Mark McCrindle tell us that modern consumers are much more engaged on an emotive scale than a cognitive scale.[24]

In his 1995 book *The Demon-Haunted World: Science as a Candle in the Dark*, Carl Sagan argued that scientific thinking is necessary to safeguard our democratic institutions and our technical civilisation. He said we need to teach both the scepticism and wonderment of scientific thought.[25]

If it was widely understood that any claim to knowledge needed adequate evidence before it could be accepted, he said, then there would be no room for pseudoscience.

So we should judge a society's scientific literacy not on what we do or don't know, but on how we think. It's not really that important if we don't know how many kilometres the sun is from the earth, or whether oxygen occurs naturally in the air we breathe or is released by plants – but we do need to know how to make decisions based on evidence, not vague claims that align with our emotions. Without that we will continue to vainly argue science facts with non-science values, where facts and logic have little impact.

One last big 'but'

But – and this last one is very big but – we need to be clear that the overall purpose of understanding the drivers of beliefs in pseudoscience, or alternative beliefs, is not to do the ha-ha aren't-you-dumb thing, so common amongst sceptics. The point is to understand.

You might find it easy to state that one life philosophy is superior to another, regardless that some people get enormous purpose and meaning from theirs. But what do we make of the fact that research – yes, scientific research – shows that those who believe in, and invoke, good luck or blessings tend to have higher performance scores across a range of tests than those who don't?[26]

Or the trickier ethical question of who has the right to say that because traditional beliefs are based on superstitions, not science facts, they are therefore less valid?

For we are also wired to tend to divide people into US and THEM camps – which science communicator Mike McRae describes in his book *Tribal Science: Brains, Beliefs and Bad Ideas*, as our tribes of similar-minded beliefs.[27] But if we can rise above instinctive fears to embrace a scientific evidence-based approach to thinking, we can surely rise above instinctive tribalism and look for points of common values that allow for a complexity of world views. Many scientists are Christians – or Muslims and Hindis or Jews – or regularly consult their horoscopes – which demonstrates that you can have a co-existence of divergent beliefs.

At the other extreme, every individual on the planet has the right to be a contender for the Darwin Awards, by taking themselves out of the gene pool through absolutely stupid – sorry, alternative – beliefs and behaviours. That's evolution at work at the individual scale.

But to allow any dangerous beliefs or behaviour to be spread throughout a society, that is detrimental to the collective gene pool – and that's something we must challenge, no matter whether it is based on our instinct or our scientific reasoning.

Calling bullshit!

That people make illogical arguments to support their beliefs is nothing new, but we really need to call out those who purposefully make illogical arguments to fool others. Particularly those illogical arguments that 'feel' right and have become the tools of misinformation.

The reason false arguments work well, in many cases, is that while the world is large and complex our brains favour simple explanations. We tend to seek out ideas to confirm what we already suspect, trust people we find more appealing, and take our own individual experiences and treat it as evidence.

Some of the illogical arguments that are used that you should look out for include:

Broken logic: An example of broken logic is citing a child who got sick from a vaccination as proof that all vaccinations are dangerous and may make your child sick. Sometimes broken logic statements feel accurate, and only when really analysed do you find they are not. Sometimes they are just clumsy, like finding an error in a report on climate change and then using that to imply that everything in the report is in error. The danger is that those looking for ammunition to support a position that is already anti-climate change science, will hook onto broken logic as if it is actual logic and refuse to be convinced otherwise.

Straw man arguments: This is something that is easily knocked over, like a person made of straw, and is usually done when an opponent claims you are saying something that you actually aren't, and proceeds to demolish your supposed arguments. An example of this is Australian radio shock jock Alan Jones stating: 'And of course carbon dioxide isn't a pollutant. It's a harmless trace gas that's necessary for life.'[28]

Michael Brown, an associate professor in astronomy at Monash University in Melbourne, has said:

> Straw man climate science ignores real world complexity. Variations from year to year and place to place are assumed to undermine the case for anthropogenic climate change.[29]

False equivalence: False equivalence is usually made by comparing something morally and emotionally outrageous to some aspect of science that someone is attacking. One of the worst cases of false equivalence I have heard in recent years is when the Australian Vaccination Sceptics Network in 2015 compared infant vaccines to rape.

Case study of one: This is usually preceded by a statement like, 'Well, when my aunty visited the Great Barrier Reef she said she didn't see any coral bleaching …', although it is more often based on a person's own single experience of an incident, such as the climate change denying Australian politician, Pauline Hanson, stating that the Reef was clearly in good condition because it looked okay when she visited one tiny part of it.[30]

John Cook, from George Mason University, has described the illogical fallacies often used by science deniers as FLICC. This stands for:

- fake experts
- logical fallacies
- impossible expectations
- cherry picking
- conspiracy theories.[31]

So keep an eye out for them, whether the argument is being made about same sex marriage or science – and don't be shy of calling 'illogical fallacy' (or just plain 'bullshit') on them.

What to do with what you now know

If you encounter attitudes that you don't agree with, rather than denigrate them try and find the belief systems that underlie those attitudes, then have a conversation about those beliefs (not the attitudes).

There are very wide ranging beliefs out there in the world, and not all of them are based on science, and not many of them are easily changed. But once you understand someone's beliefs, you will better understand where they are coming from.

Then – and this is the really important bit – try and find points of unity between you. Look for any common beliefs or values. For instance, if arguing about climate change with somebody, dig a bit deeper and see if you can both agree on the need to ensure resources are available for the future, or food security, or changed weather patterns.

Once your find something that you can both agree on you have the foundation for continued dialogue. It might not start out as a science dialogue, such as ensuring the safety of children or economic stability, but it will probably get to science before too long.

For instance, if you are confronting an anti-vaccination advocate, you can argue the merits of vaccination or not endlessly with no real outcome. Instead start discussing how to protect children in general from infectious diseases. That will soon lead you to talking about exposure to diseases. And from there you can discuss the need to keep high percentages of children immune to diseases to protect individuals. The step of logic needed is to point out that a child who is not vaccinated is only safe if surrounded by children who are vaccinated, rather than by other children who are not.

You might never get that single person to change their mind – but if you can get them to accept that others vaccinating their children is a good thing for them, and also that for them to try and convert others to be opposed to vaccination is actually dangerous to their children, that might be a realistic outcome.

Key summary points

- People have all kinds of beliefs and they define who they are.

- Many beliefs are anti-science, and scientific data does nothing to influence them.

- We are hard wired to use mental shortcuts to understand complex information, but those mental shortcuts rely heavily on our biases and intuition, and often result in incorrect assumptions.

- Mistaken beliefs can be become very ingrained and become rusted on beliefs – such as infant vaccines causing autism.

- We can more easily spot errors in other's people's thinking than we can ever spot it in ourselves.

18

The risky business of communicating risk

'Improper risk communication is emerging as an environmental risk in and of itself.'

— Kristine Uhlman, environmental risk communicator

There are some times when you will find that everything you have learned about science communication suddenly doesn't work – like you have confronted the Borg in *Star Trek* who have adapted to your arsenal, and your strategies no longer have any effect.

One of these times is when you are dealing with risk communication.

I was recently asked to come and help a government agency who was working with remediating contaminated land. They said they had a problem and I might be able to help them. So I showed up and was signed into the building and led into a meeting room full of concerned faces in suits.

I sat down, feeling a bit underdressed. I asked the concerned faces around the room to outline the problem and they described to me that the even-more-concerned-than-them communities they were working with were not being responsive to their communication campaigns, like all their communication tools were broken or something.

So I leaned a little forward in my seat, placed my interlinked fingers on the table, and said, 'You are in a risk communication environment. Standard communication will no longer work well for you' (using the tone of voice you use to tell someone they have a serious sexually transmitted disease or have suddenly been transported to a far off quadrant of the galaxy where there is no Netflix).

Defining risk communication environments

So what are risk communication environments exactly? Well, they tend to be situations where you have:

- outrage
- low trust

- high perception of risk
- high circulation of alternative reports or different positions on the science that are getting a lot of traction with your audience
- alien spacecraft that are immune to your weapons (yeah, that's a metaphor, but you knew that, right?).

Like trying to achieve behaviour change or attitude change, effective risk communication does not come easily nor quickly. But on the plus side, there is a significant body of research from people who try and impress people at parties by introducing themselves as risk communicators. And they have studied how different people can have different perspectives of risk, and how to best communicate risk to different people.

They have also found that as many attitudes towards risk tend to be based on individual values, understanding values helps inform risk communication activities.

Good risk communication is also based around partnering with stakeholders and communities to develop a shared understanding of risk and shared decisions on how best to mitigate any risks. This entails thinking about communities as partners, which can sometimes be a challenge for many agencies who are not good at relinquishing control to alien species (I think you get the metaphor thing by now).

Understanding your audience by concern and affect

A good tool for working in risk communication is to graph your audience by how much they perceive they are affected by the risk and how much they are concerned by the risk, as is shown in Fig. 18.1.

This can give you a quick summary of the type of communication or engagement you will need.

Quadrant 1. Low concern and low levels of being affected. The target audiences who are in this quadrant will require information-based approaches, such as letters, fact sheets, Q&As and face-to-face conversations. This is the easiest quadrant to work with and ideally much of your risk communication work will be aimed at moving the other quadrants into this quadrant.

Quadrant 2. High concern and low levels of being affected. People in this quadrant have high concerns about risk but are not directly affected by it. The best risk communication strategy for this group is listening to their concerns – allowing them to express them and acknowledging that they have such concerns. After that, they are most likely to be more receptive to providing them with information and working to get them into quadrant 1.

Quadrant 3. Low concern but high levels of being affected. For people in this quadrant, who are not concerned about actual risks, the style of risk communication should be education-based. You will need to use deeper levels of information sharing, conversations and going into detail to explain the risks and how they are being managed, and importantly checking that your audience understands what they are being told.

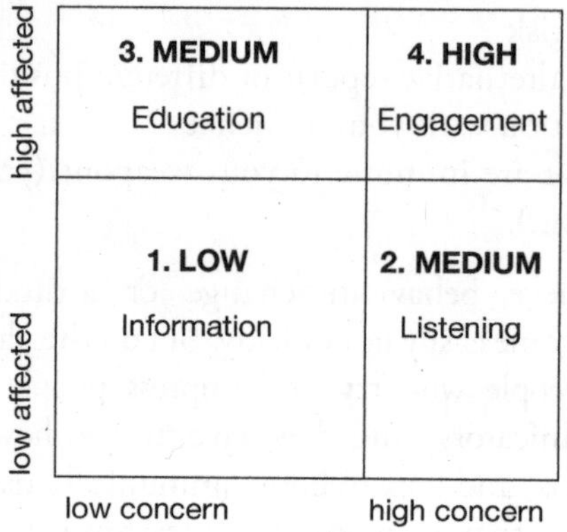

	low concern	high concern
high affected	**3. MEDIUM** Education	**4. HIGH** Engagement
low affected	**1. LOW** Information	**2. MEDIUM** Listening

Fig. 18.1. Risk communication audiences by concern and affect

Quadrant 4. High concern and highly affected. People in this quadrant can be very resource-intensive to manage, and you will need to go beyond information and education and take part in sharing of information and solutions so people feel involved in the processes. And you need to establish good working relationships with them, as relationships are key when communicating to audience with high concerns who are highly affected. This is the group you need to get risk communication right with, as you often won't get a second chance.

And never go out into the rain without packing some basic risk communication principles:

- Managing risk perceptions is not about explaining the data, but about reducing the outrage.
- Equity and control issues underlie most risk controversies.
- Risk communication is easier when emotions are acknowledged and legitimatised.
- Risk decisions are better when the public shares the power.
- Risks are best addressed by actions rather than talking.
- The public/s don't care what you know – they want to know that you care.

Traditionally when dealing with an issue of risk communication – like an environmental contamination, a food or health safety issue, or anything at all that impacts cute animals or children – someone decides to wheel out a scientist to stand in front of the outraged public and explain the data to them that shows that the risk is nowhere near as risky as they might think it is.

Like the Chief Scientist of a random nuclear power plant that is going to be built near a community, standing up and stating that the data shows that only one in every 500 000 people will die from the extra radiation being produced, so don't worry about it. Or the researcher who tells mothers that the pollution levels from the factory, or chemical spill, or pesticides or whatever, are so low that even if their children are ingesting them, they should not expect it to cause any serious health effects.

This is usually accompanied by lots of data and graphs to prove the point.

And what happens?

It rarely works. The scientists and their bosses might go home and say to each other, 'That went really well, don't you think?' 'Yeah, really well. That taught everybody what our research shows.'

And then outrage actually increases!

What the … ?

Go and read the sixth principle again. **People don't care what you know – they want to know that you care!** Risk communication requires you to adopt a significantly different mindset, in which the rules of communication are not based on the evidence of risk, but more on 'a range of evidentiary, cultural and economic considerations'.[1]

There's a good chance that most people came to the meeting not to hear you talk about the data, but so you could hear what they had to say about how they felt about the issue. So save your data for when people actually ask for it, and don't presume people are otherwise interested in hearing it.

Talking risk communication is really about listening.

Scientific perceptions of risk versus public perceptions of risk

It is important to know that when you graph the gaps between science-based thinking and a general community's way of thinking, the scientific view is often a minority one that is quite removed from the average perspectives of the general public. This means that the ability for each group to understand each other's perspectives can be difficult.

Very few members of a community are going to have a scientific view of risk. They are more likely to have a more emotional view of risk. So it is important to know that rather than trying to correct people's attitudes to how you think, it is more effective to work with them. As we've hopefully learned already, people with strongly-held beliefs will not easily abandon those beliefs – even when shown contrary evidence.

Discussions about risk need to take into account the different perspectives of risk that can exist, and to have a genuine conversation these views of risk need to be acknowledged. According to two of the gurus of risk communication, Vince Covello and Peter Sandman, a community perception of risk can be best understood as a 'threat, either real or perceived, and either quantified or non-quantified, to that which we value'.[2]

Elements that escalate that threat to things we value include:

- questions of who is to blame for the risk
- alleged secrets and perceived attempted cover-ups
- human interest through identifiable heroes, villains, victims, etc.
- links with existing high-profile issues or personalities

Scientific assessments of risk can differ greatly from public assessments of risk

Table 18.1. Fright factors that influence perceptions of risk[4]

Things are perceived to be low risk if they are seen as:	However, things are seen as more risky if they are:
Voluntary	Involuntary
Common	Rare
Not fatal	Fatal
Known	Unknown
Known to science	Not known to science
Controllable	Not controllable
Old	New
Immediate	Delayed
Natural	Artificial
Detectable	Undetectable

- exposure of many people ('It could be you!')
- strong visual impact.[3]

Fright factors that influence perceptions of risk are summarised in Table 18.1.

Hazard *v.* outrage

When working in a risk communication environment it is also important to understand the differences between hazard and outrage. According to Peter Sandman, hazard can be understood as how much harm something is likely to cause, while outrage is how upset it is likely to make people. For instance, the chemicals that people voluntarily expose themselves to in hair salons are actual hazards (but have low concerns), while the low-risk nanoparticles in sunscreens cause outrage (but are low levels of hazard).

Scientists tend to concentrate on the hazard and miss the outrage.

Peter Sandman has suggested that risk should be understood as:

$$risk = hazard \times outrage[5]$$

He has also stated that in risk communication environments you should respond to hazard when actual risk is high, and respond to outrage when perception of risk is high. So:

1. When **hazard is low, but outrage is high**, the task should be 'outrage management' – reassuring excessively upset people. This is the most common situation for risk communication, when there has been a sound policy decision based on good science, but nevertheless people are concerned or upset about it – such as putting fluoride into public drinking water. The incorrect default solution is to try and explain the science rather than address the outrage or concerns.

2. When **hazard is high and outrage is low**, the task is 'precaution advocacy' – alerting insufficiently upset people about serious risks. There are usually some

who think it better to not talk about the elephantine gorilla in the room, and avoid talking about the hazard, but that is short-term gain for long-term pain. When the outrage inevitably starts building, you will have forfeited a position of trust that could have been established from providing early precautionary advice.

3. When **hazard is high and outrage is also high**, the task becomes 'crisis communication' – helping appropriately upset people cope with serious risks. It requires an approach of 'We'll get through this together.'[6] And crisis communication is actually a whole other subset of communication, better handled through engagement (see Chapter 14) than standard communication practices. Acknowledging uncertainty is also important, as is transparency, laying down trust to draw on later.

There is also the often unstated outrage that should not be ignored: the outrage by experts and senior managers that the public are not listening to their data, and the outrage at the cost of having to undertake some form of engagement with the public. In such cases the best thing to do is to put on an alien Borg suit and tell them, 'Resistance is futile'.

Cognitive biases: Understanding how people think

Many of the problems relating to risk communication stem from the fact that scientific definitions of risk can be very different from community perceptions of risk. Risk perceptions are largely driven by the ways we think, often overestimating or underestimating risk, and making emotionally-driven decisions. For instance, people perceive flying to be more risky than it actually is and perceive driving to be less risky than it actually is. Or people perceive texting while driving to be more risky for other people than themselves, or they perceive spending too much time in the sun without sunscreen as risky for their kids, but not themselves.

The heart of the problem is the way we are wired psychologically, which leads us to common errors in our thinking that can lead to distortions of perception, inaccurate judgments or illogical interpretations of risk. Ask a smoker to tell you how risky they think it is to smoke. Or ask a teenager how risky it is to drive fast. Or ask a mum at the organic shop how risky it is to eat GM foods.

Common cognitive biases that influence the way we think (which are addressed in more detail in Chapter 17, but are worth revisiting), include:

*   When we are time poor, overwhelmed with data, uncertain, driven by fear or emotion, we tend to assess information on mental shortcuts or VALUES not FACTS.
*   Most people, when faced with an issue related to science and technology, adopt an initial position of support or opposition, based on a variety of mental shortcuts and predisposed beliefs rather than scientific evidence, such as:
    *   an anthropocentric world view favours climate change denial
    *   a natural values world view favours an anti-GM stance
    *   a right to life world view favours anti-embryonic stem cell research.

- When people are shown information proving that their beliefs are wrong, they can actually become more entrenched in their original beliefs.
- The more people with opposing points of view talk about the topic, the less likely they will agree on any issue or even see it the same way.

For a more detailed list, and a giant infographic on cognitive biases, check out Buster Benson's 'Cognitive bias cheat sheet'.[7]

This means that your scientific data on risk is not just likely to be ineffective, it is likely to backfire on you and cause more concern and outrage.

Risk perception

Risk perception underpins a lot of risk communication work, but it can a problematic term to use, as it implies that a particular view of risk is only a perception, or a mistaken view of risk that needs to be corrected. It is better to know that perceptions become realities for the people who hold them, so to them they are not risk perceptions, but actual risks.

In short, opinions that were NOT formed by LOGIC or FACTS are not able to be easily influenced by LOGIC or FACTS.

When people are stressed, their perceptions and decisions are influenced by a wide range of factors, with technical facts often being the least important (worth less than 5%). It is important to understand that people cannot focus on facts when there is any level of outrage.

Another phenomenon that increases the perception of risk is known as the social amplification of risk, whereby certain things can cause 'ripple effects', like a stone being thrown into a pond (or a Borg photon laser being ignited in near-earth orbit). These can increase both the intensity and longevity of a perceived risk. One common ripple effect is 'stigmatisation', which is the ongoing association of a risk with negative qualities and high risk.[8] For example, a media story about the risk of swimming during a shark migration will lead to people staying away from the water for a long time after the sharks have swum on by.

Trust

Trust is also an important part of effective risk communication. As we discussed in Chapter 13, when a spokesperson or expert stands up in front of any audience, their trust is assessed in the first 15 seconds or so. It is often based on perceptions of being empathetic and caring – which far outweigh any perceptions of expertise and knowledge.

To build trust with a community, it is more important that you listen to them rather than speak to them, as people want to tell you what they feel, before they hear from you. And when there is a lot of uncertainty, acknowledging what is as yet unknown and uncertain about a risk can build community trust.

Empathic risk communication also helps build trust. It aims to get your stakeholders' feelings 'into the room' without making your stakeholders feel exposed. This is based on acknowledging what people feel and validating that it is an understandable way to feel.[10]

Explaining risk

It is important to know that how people understand calculations of risk can influence their understanding of risk. And using only words leads to a great variation in understandings. For instance, telling someone they have a low chance of experiencing a side-effect of a medicine, is generally ineffective, and can actually increases the perception of risk.

It is more effective to provide numerical estimates of risk, but even then it must be done with care. The US FDA recommends a 10-point rule for communicating risk:

1. Provide numeric likelihoods of risks and benefits.
2. Provide absolute risks, not just relative risks.
   Focus on the individual's risk – not an individual's risk relative to others.
3. Keep denominators constant for comparisons.
   For example, 1 in 10 000, 337 in 10 000.
4. Keep time frames constant.
5. Use pictographs and other visual aids when possible.
   Graphs make numeric information easier to understand and infographics can be the best way to communicate knowledge.
6. Make the differences between baseline and mitigation of risk clear.
   Use pictographs or infographics to show baseline risks in one colour and the risks due to treatment in a different colour.
7. Reduce the amount of information shown as much as possible.
8. Provide both positive and negative frames.
   People, particularly those who are less numerate, are unduly influenced by whether an action is described in positive or negative terms (e.g. survival rates versus mortality rates). Whenever possible, describe the risks and benefits using both frames.
9. Take care using interpretive labels or symbols to convey the meaning of important information.
   Interpreting the meaning of numeric information (in terms of its goodness or badness) can affect people's risk perceptions and change their decision making, so make sure you use symbols that work for your audience, rather than work for you.
10. Test communications before use.[9]

Sandman also says that being able to 'deflect' what you say leads to less likelihood of provoking outrage. He cites an example where there has been environmental pollution, and people will more easily accept that there could be some concern about property values if you raise it, rather than admitting that they themselves are actually more worried about property values than, say, public health (see Table 18.2).

Media

A word on working with the media, for despite all your best efforts everything can come crashing down around you – not unlike a Galaxy Class battle cruiser under fire from a Klingon Bird of Prey – if you get a run of bad media coverage. And yes, working with the media in a risk communication situation can be a risk in itself – but sometimes it can be a bigger risk not to work with the media.

Table 18.2. Deflecting statements[10]

Undeflected	You	'You're not really worried about health! You're afraid your property values might be affected.'
Deflected	I	'I was in a situation like this when I lived near an industrial park. What worried me even more than the health effects was the possibility that my property values might be affected.'
More deflected	They	'One of your neighbors was talking with me last week about this situation, and the thing that worried him the most was the possibility of an effect on his property values.'
Still more deflected	Some people	'Some people in a situation like this would probably be worried about their property values.'
Most deflected	It	'It's possible there could be some concern about property values here.'

Understanding the nature of the media in risk situations can help a lot:

- In risk stories reporters tend to cover viewpoints, not 'truths'.
- The risk story is usually simplified to a dichotomy.
- Journalists are more interested in simplicity than complexity.
- Journalists try to personalise the risk story.
- Claims of risk are usually more newsworthy than claims of safety.
- Journalists do their jobs with limited expertise and time.[4]

It is just a fact of the world we operate in that for many risk scenarios, minority voices claiming there is a particular risk are likely to get disproportionately higher media coverage than any scientific experts who are claiming that risks are actually much lower.[11] And uncertainty can be exploited to imply that uncertainty equates to high risk.

Added to this, the media's propensity to try and get balance to a story means they will seek a dissenting voice – even a scientific voice – and give it near-equal air time regardless of how much outside the scientific consensus it is. Often, they are not even from the discipline under examination. Look at the number of scientists who are dissenters about the causes of climate change, and check their scientific fields of expertise.

The US Centers for Disease Control has an excellent booklet on working with the media in a crisis that says it is important to remember that it is a journalist's job to provide balance by looking for alternative perspectives and interpretations of events, and ensuring that other points of view receive coverage. You need to be aware that your side of the story is not the only one that will get picked up and that you can only expect limited success in influencing any media coverage devoted to debate, discussion, and speculation.[12]

However, it helps your case if you make things as easy as possible for journalists to do their job, so make your points clear, concise and consistent.

And if you find the media have presented incorrect information, especially if it could be harmful to members of the public, quickly communicate the correct information to both the media, and the public through your own social media channels.

Another to-do list for working with the media, provides 10 key points:

1. Be open with, and accessible to, reporters.
2. Respect their deadlines.
3. Provide information tailored to the needs of each type of media, such as sound bites, graphics and other visual aids for television.
4. Agree with the reporter in advance about the specific topic of the interview and stick to the topic in the interview.
5. Prepare a limited number of positive key messages in advance and repeat the messages several times during the interview.
6. Provide background material on complex risk issues.
7. Do not speculate – say only those things that you are willing to have repeated.
8. Keep interviews short.
9. Follow up on stories with praise or criticism, as warranted.
10. Try to establish long-term relationships of trust with specific editors and reporters.[2]

What to do with what you now know

If you find yourself in a risk communication situation, and that your general communication tools no longer seem to be working, reach into your risk communication tool box.

You will first need to have a close look at the three basic principles of knowing your audience, having a clear message and knowing why you are communicating. Odds are you will find that you didn't know your audience as well as you thought, didn't have a message that was closely enough focused or clearly articulated and you might even have to rethink your objective for communicating. Be very clear if your communication objective and message are about 'evacuate now', or 'let me help you with the uncertainty', or even 'validate your feelings', or some other key need.

For instance, when communicating with someone who does not support infant vaccination, the objective isn't really about getting them to change their mind as an individual, it is about ensuring enough children in society are vaccinated so there is effective herd immunity. A subtle but important distinction.

The next thing you will need to bring out of your tool box is the idea that throwing information at people in a crisis situation is unlikely to achieve useful goals. This is mainly because crisis situations are about emotion, and any form of communication that doesn't start with acknowledging emotion (and also end with acknowledging emotion, and connect every component to emotion) is going to have a problem.

A useful tool in any risk or crisis situation is to be very open in telling what you know and what you don't know. Being certain about uncertainty is very important.

Another good tool to bring out of your risk communication tool kit is the fact that very few members of a general community are going to have a scientific view of risk. Again, their view of risk is much more likely to be an emotional one.

We all use both emotion and thinking to respond to situations, but in a high perceived risk situation, emotions will dominate.

Social media

Social media often plays a large role in risk communication, with individuals or groups setting up accounts that collate scare stories and rare data that emphasises a risk where your scientific data might say there is only a small risk. If this is happening, again you need to know that holding up your data in defence of a social media story or campaign can have little effect.

Social media, particularly in risk situations, is not about discussing facts and figures. It is much more about experiences and narratives – or the triumph of the experiential and emotional over rational thought.[13] So social media in risk communication is best viewed as an extension of community conversations, and a good source of understanding what communities think and feel. It should rarely be considered another communication channel – especially when attempting to counter incorrect information being posted on social media. As this is generally done from positions of low trust, its impact will be limited.

On the plus side social media can provide a rapid way of connecting communities and individuals and it can provide instructions, or reassurance, rather than just spreading fears and misinformation.[14]

Key summary points

- The normal communication rules don't often apply in risk communication scenarios.

- People tend to overestimate or underestimate risk, and will invariably have a different view of risk than the scientific assessment of risk.

- It is more important to deal with outrage than to explain the hazards.

- Attitudes that were not formed by facts and logic will not be influenced by facts and logic, so put away the data until it is explicitly requested.

- The public don't care what you know – they want to know that you care.

19

Valuing values

Cecil Graham: 'What is a cynic?'
Lord Darlington: 'A man who knows the price of everything, and the value of nothing.'

— Oscar Wilde, *Lady Windermere's Fan*

This is going to be a fairly short chapter, as much of the content is scattered throughout the rest of the book, and you should have picked up by now the importance that understanding values plays in understanding how to better communicate with people. As we have seen, our values not only define who we are, but also define how we think. When science and technologies challenge people's beliefs and threaten their deeply held values, their attitudes towards science and scientific information can be strongly affected, leading them to deny or dismiss the validity of the science.

But first a story.

I was giving a seminar to a group of communicators who work on trying to communicate the science of bushfires to the public, and I said, 'You know how the recent appearance of more frequent wildfires has changed the rules on everything?'

Lots of heads in the room nodded. They knew that wildfires have become so huge and so strong and so devastating that they can kill even the microbes in the soil, and that the best containment lines and clearing will do little to even slow their terrible pace. They knew that wildfires can create their own weather patterns, funnelling smoke and particulates high into the stratosphere, even producing powerful lightning strikes that trigger new fires miles away from the fire front.

'Well,' I told them. 'It's the same for modern communication. We are facing a perfect storm of change that creates its own weather patterns, and the tools we have traditionally used don't work so well anymore.'

That perfect storm for communication includes an abandoning of traditional media outlets, the self-selecting echo chamber effect of social media, the rise of the passionate individual as expert, the popularisation and polarisation of politics, and the empowerment of many of our individual biases as a result of these – based on our values.

Let's be quite honest here – we all use our values to make decisions. This is not something that scientists and science communicators are immune to. We have no more inbuilt natural immunity to the rule of bias over reason due to our righteousness than vaccine deniers have immunity to measles and whooping cough due to their righteousness.

But understanding our values and recognising when they are coming into play can help us understand how to manage them.

I took part in an exercise to go into small communities around the Australian state of Victoria to talk to people about their attitudes towards fires, and to find out what they could do as a community to minimise their risk of fire. It quickly became apparent that the communities were not particularly interested in all the science and data on fires, but were more interested in the things they valued in their communities that fires threatened. So we created values maps, based on things like, *Community feeling, Wilderness, Closeness to nature, Our local school* and so on.

When it came time for the fire authorities to talk to similar communities about fire risk, we advised them to put away their statistics and charts and instead hold conversations with the communities about the things they valued. Then they would have an opening to have conversations about how to protect those things from risks such as fire.

Misunderstanding values

Likewise, biologist Andrew Thaler has said on his blog *Southern Fried Science*, that when he has to talk to an audience about climate change, he starts by talking about things that are important to his audience. 'Such as fishing, flooding, farming, faith and the future.'[1] And having established that they become an entry into discussing global warming. As such the audience is able to see scientific evidence as relevant to their values, not contradictory to it.

The values prism: a tool to understand values and how to approach them

We've talked a lot in previous chapters about how to measure and understand the values of an individual, or a group or a community, through dialogues or surveys or engagement processes, but that is only half the task. You need to then compare those values with your own (or those of the agency you work for). After all, every data point is understood by its relation to another data point.

If you can map the differences in values, you will have a firm idea where there will be the closest alignment of values and where things will work relatively smoothly, or where you will have the largest values clashes that will likely impede effective communication.

The values prism, shown in Figs 19.1, 19.2 and 19.3, is a useful tool for mapping these differences. It works like this. You draw up a values prism, based on the spider web chart or radar chart. You can put in as many or as few dissecting lines as you need, giving each the name of a particular value that you see as being relevant to the communication task you are working on. They might be *trust in government*, or *independence*, or *personal health*, or *protect the environment*. For users of national parks, they might be *altruism, egoistic and hedonistic* values.

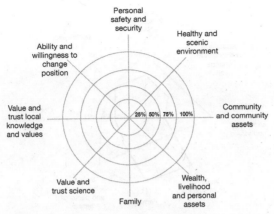

Fig. 19.1. The values prism

Label the very centre of the chart as 0% and each ring afterwards as 25%, 50%, 75% and then 100%. Then you can map the values of the community or group you are working with.

I am sometimes asked, 'But what if you don't know their values?'

Well find out. You may even ask them to map their values themselves.

And, yes, I know you will sometimes get a huge range of responses from a community, with lots of different values across any single issue, but then you just map the intensity across issues, or look for the most common responses.

Once you've done that you will have a chart that gives you an approximation of values. It's a rough tool, but one that provides enormous insights.

Then chart your own values, or those of the agency you work for. Where there are points of alignment, communication

There is nothing quite like asking the right question to get the answers you need.

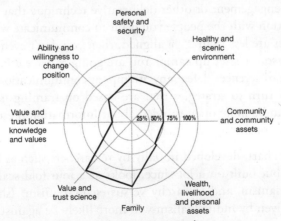

Fig. 19.2. The values prism with one values set mapped

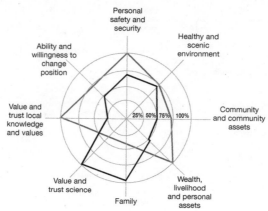

Fig. 19.3. The values prism with two values sets mapped

and engagement is probably going to be fairly smooth. But where there are significant gaps, you know you are going to have to work a lot harder.

But wait, there's more! (as they say on late night infomercials on TV, trying to sell you crap when you're drunk and watching re-runs of *Bewitched* or *The A-Team* – or so friends tell me!). The size of the values gaps guides you in knowing what types of communication strategies are likely to be the best to use:

- If your values are 75% to 100% aligned, you are both pretty much in the same camp on an issue and you won't have to slug it out to get your message across. You can expect to be able to use more information-based strategies.
- If your values are 50% to 75% aligned, you will be a little bit in the same camp, but only a little bit, so will probably need to be using more education-based strategies, that explain things in deeper levels of detail, and have people participate in more active learning.
- If your values are only 25% to 50% aligned, you aren't in the same camp, but can see each other's tents in the distance. You are likely to need to use some form of community engagement or other deliberative technique that involves discussion and deliberation with the people you want to communicate with.
- If your values are less than 25% aligned, then you don't even really know where the other person has set up camp. You are going to have a lot of challenges and may find the differences defeat any attempt at communication. In such cases you may need to turn to strategies that are based on framing (see Chapter 12), as people tend to be more open-minded about information presented in a way that appears to be consistent with their values.[2]

Another values chart, developed initially by researchers such as Mary Douglas and Aaron Wildavsky[3] but built on a lot since, looks at a four-fold scale of individualism versus communitarianism, and hierarchy versus egalitarianism[4] (shown in Fig. 19.5). Someone who is driven by individualism will more likely be against infant vaccination

Things to know about values

- When information is complex, people make decisions based on their values and beliefs.
- People seek affirmation of their attitudes (or beliefs) – no matter how fringe – and will reject any information or facts that are counter to their attitudes (or beliefs).
- Attitudes that were not formed by logic are not influenced by logical arguments.
- Public concerns about contentious science or technologies are almost never about the science – and scientific information therefore does little to influence those concerns.
- People most trust those whose values they feel mirror their own.

as their decision to not vaccinate is overwhelmingly based on personal risk to their child, where as someone who is more communitarian will most likely agree to have their child vaccinated as it is something that is good for the whole society.

Fig. 19.4 shows that someone who lies in the hierarchy-individualism quadrant will more likely see climate change, nuclear power and guns as a low risk, but will see gun control and immigration as a high risk. In the opposite quadrant, people who are egalitarian-communitarians will more likely see climate change, nuclear power and guns as a high risk and gun control and immigration as a low risk.

Those who lie in the hierarchy-communitarian quadrant will more likely see the legalisation of marijuana and gays in the military as a high risk, but see anti-terrorism actions as a low risk. Those on the opposing side, in the egalitarian-individualism quadrant will more likely see the legalisation of marijuana and gays in the military as a low risk, but see anti-terrorism actions as a high risk.

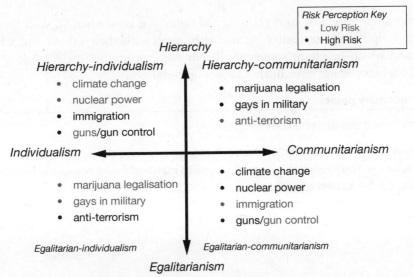

Fig. 19.4. Cultural cognition world views[5]

What to do with what you now know

Values can be the key to understanding a huge amount about your target audience or stakeholders. Understanding those values gives you the capacity to align information with those values. Like beliefs, values don't change easily, and underpin attitudes and behaviours in many instances.

Working out people's values isn't as hard as it might sound, because people are usually very happy to talk about their values and the things that they value.

So values-based conversations are a very good way to start engagements of any kind, as they will not only give you some of the insights you need, but will suggest starting points for deeper levels of engagement.

Use a values prism to map your own and your audience's values, if you need to. It is a useful tool to help you simply get a grasp of values gaps that may exist. Then, start developing strategies and messages that align with your audience's values.

If you find that your values are irreconcilable, there are other ways to effectively communicate a point. For instance, when I'm confronted with somebody who strongly does not believe in climate change as it threatens their fundamental values of, say, economic progress and development and life-style choices, I say:

'You don't have to believe in climate change to accept that the world is getting warmer. You don't have to believe in climate change to accept that we are seeing more extreme weather events like droughts and storms. And you don't have to believe in climate change to know that we need to spend resources on mitigating these events.'

In other words, don't debate whether climate change is human-caused or a natural cycle in the world – that isn't going to get you anywhere with some people. Instead, discuss the types of mitigation that will be needed to deal with the impacts of changed weather events. That is something that is going to be a constant despite what you believe in.

This can go some way to explaining the fights you have when all your family gets together. People differ on issues because they clash with their values, and when you understand which quadrant somebody lies in you can predict their values a bit and know the topics you are most likely to agree or disagree on.

Key summary points

- It's a short chapter, go and read it all!

- Being able to map different values between yourself and the people you are seeking to communicate with enables you to match different communication approaches to best effect.

20

Can you really change behaviours?

'God grant me the serenity to accept the things I cannot change, the courage
to change the things I can, and the wisdom to know the difference.'

– Reinhold Niebuhr, ethicist and philosopher

Let's jump straight to the crux of the issue here without any framing anecdotes or stories:
is it actually possible to change people's behaviours in a sustained way? Anyone with
children – especially teenagers – is going to have their own thoughts on that. But the
research clearly says yes.

Well mostly yes.

Well mostly yes in some situations.

Well mostly yes in some situations, but maybe not in others.

But if there is one thing that researchers who are studying behaviour change all
seem to agree on, it is that there is little agreement on how to achieve and sustain
behaviour change. So much so, that a report by the European Food Information Council
identified over 60 different socio-psychological models and theories of behaviour
change. The key ones include:

- *Self-determination theory*, which states that individuals must really believe the
  behaviour is enjoyable or compatible with their sense of self, their values and their
  life goals, if they are to undertake the behaviour change.
- *Social marketing*, which states that smaller initial behaviour changes are crucial to
  achieving any larger behaviour changes.
- *Nudging*, which makes the desired behaviours easier options to adopt.[1]

But one of the things that researchers do agree on, in regard to behaviour change, is
that you don't get people to change their behaviours by first changing their attitudes –
even though that feels the intuitive way to go. It is actually more achievable to get people
to change their behaviours which will lead to a change of attitude. Campaigns based
around increased sunscreen use or anti-littering take this approach. Make a behaviour easy
or more socially acceptable, and people change their behaviour and that then leads to an
attitude change in line with that behaviour, which will be more likely to reinforce it.

This is because of a human trait known as cognitive dissonance, which means you
cannot easily have a behaviour and an attitude that clash with each other, and you tend
to change your attitude to align with your new behaviour. For instance, have you ever

had a friend or family member who was really anti-big business as a student and then got a job in a big firm, and you found that they changed their position and started defending big business?

That's cognitive dissonance in action!

Just get people to act

Consumer psychologist Adam Ferrier says it this way: 'Action changes attitude faster than attitude changes action.' He also says that when developing messages to achieve behaviour change, you should forget rational messaging and also forget trying to use advertising or communication to forge an emotional connection to people, for you will get greater behavioural change if you just get people to act.[2]

However, these actions need to be simple and easily achieved, and possess a motivator. He calls this a 'spur'. A key spur is ownership. If you can give ownership of the action to people, they will value it more and will be more likely to change their behaviour in accord with it. Some useful psychological ownership principles are:

- **The hawthorn effect:** This effect is demonstrated by simply asking someone's opinion on a topic, and as a result they tend to value that topic (or action) more.
- **The endowment effect:** This effect states that if people physically hold something, even for a few seconds they'll value it more.
- **The Ikea effect:** This effect shows that people value something more if they co-created it.[2]
- **The Lake Woebegone effect:** This effect, named after the radio show by Garrison Keillor ('Where all the women are strong, all the men are good looking and all the children are above average'), shows that we tend to overestimate our own abilities to do something, and if you flatter people by telling them they will do better at something than most, they will be more keen and generally will do it well.[3]

Nudge, nudge, wink, wink

Nudging has become the go-to strategy for behaviour change in recent years, and has been widely adopted by the US and UK governments in large-scale campaigns aimed at achieving desired behaviour change for public benefits.

Some examples of nudging include:

- In 2009 authorities at Schiphol Airport in Amsterdam installed small fly-shaped stickers in the men's urinals and found that having something to aim at reduced spillage by 80%.
- The American grocery store Pay & Save placed green arrows on the floor that led to the fruit and veggie section of the store and found that most shoppers followed the arrows and increased their purchasing of fresh produce.
- Restaurants often have something on the menu that is ridiculously expensive. This is there so that the other things don't seem so expensive in relation to it.

Behaviour change theories and models

Health education studies have looked at different approaches to behaviour change and have developed several models, including:

Bottom-up participative strategies: These focus on empowering and resourcing local groups and networks to identify problems, define solutions and initiate action plans of their own. Examples include Community Fire Units, and the American Red Cross's Disaster Resistant Neighbourhood program.

Health belief model: The person must believe that there is a significant risk to them and the suggested benefits will compensate for the cost of undertaking the necessary behaviour.

Social cognition theory: Importance is placed on an individual feeling that they have some sense of control over their behaviour and the outcomes they want to achieve.

Behaviour commitment: Individuals with stronger commitment to a behaviour change maintained behaviours at a statistically significant level, even a year after the study period, and even recruited other people to serve as agents of change.[6]

- In the UK, people who were behind in paying their taxes were sent letters that told them things such as nine out of 10 people in your area are up to date with their tax payments, which led to an increase in keeping up to date on payments.[4]

Some of the basic features of such campaigns involve:

- examining all the applicable theories in the literature
- trialling and re-trialling methods until there is confidence that they will work
- acknowledging that messages that appeal to some people's values to affect behaviour change may have no impact on those with different values.[5]

But of course, there are limits to what behaviour changes are possible to achieve through communication campaigns. For instance, in the field of public health, huge amounts of resources have gone into trying to make people increase healthy choices about diet or exercising more, or smoking and drinking alcohol less, with limited success. But they have helped provide data on what works and what doesn't.

The complexity of our behaviours

Few of our behaviours are fully conscious, and most are just learned habits that become routines. So the challenge for a communicator is not just to change a behaviour, but to change a routine that exists around it. This means that successfully changing behaviours is not only about changing one act; it is about altering the routines in which the acts are embedded.[7] For instance, if you wanted to change someone's drinking behaviour, you would have to change the routine that takes them to the bar each evening, or puts them in front of the television with a can of beer and so on, by finding an alternative behaviour.

Researchers have also found that behaviour change is more effective when it focuses not just on the behavioural outcomes, but the steps required to reach those outcomes, as complex actions can be made up of several discernible behaviours.[8]

There are many taxonomies, or analyses of different types of behaviours, which can be divided into four main behaviour types:

- impulse behaviours
- routine behaviours
- causal behaviours
- thoughtful behaviours.[9]

Of these, thoughtful behaviours are the easiest to change, based on communicating skills transfer, knowledge and attitude.[9]

And of course, there's values

A 2011 study looking at the relationship between attitudes and behaviours in terms of water consumption found that your individual values directly influence your behaviour. For instance, if you had a high concern for the environment, as well as high water conservation awareness and practice, you were much more likely to be willing to reduce your water usage.[10] Attitudes alone, such as knowing that water conservation is important, have a much weaker link to water conservation behaviour.[11]

Likewise, it has been found that specific 'positive attitudes' towards the environment do not alone predict whether you might take specific environmental behaviours, such as buying a more fuel-efficient car. But having a specific 'concern' about the environment is much more impactful on your behaviours.[12]

However, it is important to understand that behaviours are not static, and people may continually adapt their behaviours, for many reasons. But people do tend to act in ways that are consistent with their values and beliefs.

Behavioural economics

So what is all the fuss about behavioural economics? I mean, what has economics even got to do with science communication? It's a fair question, but let me tell you a story.

Back in 2008, the world went through a serious recession known as the global financial crisis – despite most of the senior economists of the world stating how robust the global system was. The realisation that they had got it wrong was well articulated by Alan Greenspan, former chairman of the US Federal Reserve, who told the US Congress that he was 'shocked that the markets did not operate according to his lifelong expectations'.[13] Moreover, he admitted that he had 'made a mistake in presuming that the self-interest of organizations, specifically banks and others, was such that they were best capable of protecting their own shareholders'.[13]

Well – duh! But his error was in believing that most people and institutions they work for, act in rational ways.

Dan Ariely, professor of psychology and behavioural economics at Duke University in the US and author of books such as *Predictable Irrationality* has said, 'We are finally beginning to understand that irrationality is the real invisible hand that drives human decision making.'[13]

Put simply, behavioural economics uses psychology and economics to understand the cognitive biases that prevent us making rational decisions – and more importantly, how those same biases and irrationality can be used to influence behaviour changes.

Two names worth checking out, if you want to read further, are the Nobel Prize winners, Richard Thaler (author of *Nudge*[14]) and Daniel Kahneman (author of *Thinking, Fast and Slow*[15]).

Many governments around the world have established behavioural economic units, including the UK Government's Cabinet Office, the Singaporean Government and the New South Wales Government's Department of Premier and Cabinet. President Barack Obama not only established a Social and Behavioural Sciences Team when he was in office, but issued an executive order instructing federal government agencies to apply behavioural science insights to their programs.[16]

But to my mind the real strength of behavioural economics is the fact that it often relies on randomised controlled trials to determine how well an intervention actually works. Imagine if all your science communication work was tested by randomly assigning test subjects to two groups, one to test a proposed communication activity on, and the other as a control group with no intervention.

OMG – that would mean applying scientific principles to science communication! Unheard of, I know. Science communication is best done flying by the seat of your pants and making it up as you go along – not!

As an example of how behavioural economic principles can be applied, a team of researchers from CSIRO in Australia looked at how they might be applied to encourage household energy saving, which is shown in Table 20.1.

A cautionary note about behavioural economics though. It is very good for getting behaviour changes, but as it is not based on aligning behaviour with existing values, it may not maintain longer-term behaviour change if the incentives are removed. Yes, you may get attitudinal change, as we have seen behaviour change leads to – but it is less likely to impact fundamental existing values.

For instance, Adam Cormer and Alex Randall from the UK make the argument that if an organisation is seeking to promote the conservation of a forest and uses behavioural economic tactics to achieve changed behaviours that support this – that is good – it might not actually achieve more long-term pro-environment beliefs from those who were never 'true believers', even though they have changed their behaviour.[18]

So look carefully to your objective. Do you want just a behaviour change or a deeper public buy-in to a particular way of thinking?

When it doesn't work

As stated earlier, the large amount of research that has been done on trying to achieve behaviour change has been pretty good in finding what doesn't work so well. This is known as negative behaviour change – and you can think of it like nagging your spouse

Table 20.1. Applying behavioural economic principles to encourage household energy savings[17]

Behavioural economic principle	Application
1. People prefer the status quo.	Consumer acceptance of energy-related initiatives can be significantly improved by having a default option of inclusion and people must actively seek to 'opt-out'.
2. People prefer to exert only the effort needed to achieve a satisfactory result rather than an optimal result.	By making a desired action easier and more convenient it is more likely to be taken up. Conversely providing too much information can limit the uptake of a new behaviour.
3. People typically focus on the risks, costs or losses associated with adopting a new behaviour, rather than the gain. Losing something makes you about twice as unhappy as gaining the same thing makes you happy.	Energy-saving messages should be framed in terms of avoiding or minimising prospective costs and losses, not just the gains. Also highlight how energy conservation activities and pro-environmental behaviour will prevent future losses and costs.
4. People will keep on with something if they have invested time, money and energy into it.	Messages should be framed to reduce the focus on any large costs of time, effort and money that consumers have already outlaid for old energy-inefficient items, while drawing attention to the ongoing costs, losses or risks associated with retaining such inefficient items or wasteful energy practices.
5. People are motivated by rewards.	Monetary rewards, however, don't prove as consistent in achieving behavioural change as non-pecuniary rewards like praise, recognition and social approval.
6. People use trust to guide decision making when assessing risk and making cost-benefit appraisals.	Information and incentives are more likely to motivate and have a bigger behavioural impact if they come from credible, trustworthy sources.
7. We are very sensitive to the thoughts and actions of others and seek their approval.	Messages should emphasise what peers are doing and what community expects of people.
8. People draw readily on available information that is easily accessible in their memory – such as personal anecdotes of family and friends – to inform decision making.	Messages should incorporate examples of energy-saving activities that are easily available in consumers' memories, such as drawing on recent and favourable consumer testimonials. Studies have also found that simple prompts and reminders can increase a range of pro-environmental behaviours, such as energy conservation.

or kids. You probably know it doesn't work, you watch how it doesn't work every time you do it, but it just feels like it should work, and so you keep doing it.

'**Second-order conditioning**' is the term for a behaviour that has been learned in one particular context but does not work in another context. For instance, your kids might behave themselves and eat with good manners in a restaurant – but at home, eat like zombies in a brain museum.

Another example is how people have learned to separate their plastic, paper and glass in recycling when different bins for this are provided – but they have become so reliant on the bins that the behaviour fades when the different bins are not provided.

Five key behavioural insights to consider adopting

1. **Power of free.** Yes, we love the word free and it can release large quantities of dopamine into our brains – but only if we believe there was actually a higher cost involved originally, and that thing is now being offered for free.
2. **Show what others have done.** This is known as social norms. We are social creatures and respond very strongly to conformity and like to behave like we think the majority of people are behaving. An example is how hotel guests are told that the majority of hotel guests reuse their towels, which increases towel reuse.
3. **Dominated alternatives.** If there are two choices for your audience, you can steer them towards a preferred option by introducing a third option that frames your preferred option as more desirable. This is usually done by having the new option clearly inferior to the preferred option. An example is a Chinese study in which factory workers were provided with spray bottles of sanitiser to clean their hands and workspaces – and told to do it hourly but had less than optimal uptake. After measuring usage, the workers were offered another less-convenient choice: a squeeze bottle of sanitiser or a wash basin. The outcome was that use of the spray bottles increased from 60% to over 90%.
4. **Irrational value assessment.** If you are told something is very significant, or worth a lot, you are more likely to think better of it. Like being told that if you read this book you will not only be smarter – but if you are a woman you will be stronger, if you are a man you will be better looking and if you are a child you will be above average.
5. **Decision paralysis.** Having too many options can lead to a lack of decision. Dropping the number of options to about five gets a better result, as is often done with supermarket products – giving you choice, but not so much choice that it inhibits your ability to make a decision.[19]

What to do with what you now know

Let's be very honest about this. Achieving behaviour change, and sustaining it, can be very, very difficult. If you are not convinced of this, trying getting teenage kids to do something that you want them to do, but they aren't so keen on. That is the problem in a nutshell.

The huge variety of behavioural change theories doesn't necessarily help you, as it can make strategic choice difficult, and it is impractical for most people to trial dozen different theories to find the best one.

As in all things where there is a huge variety of choice of approaches, my best advice is to concentrate on the ones that are best-suited to your particular situation.

However, there are several behaviour change theories that might work better for most science communicators, as they have proven effective for getting some sort of behaviour change in areas around health and environmental behaviour.

It is also useful to look at case studies of effective behaviour change that are similar to what you are seeking, such as better health behaviours or better environmental behaviours.

And here's one to learn from. In the Australian state of Victoria there was a campaign to stop people from letting their cats wander the neighbourhood at nights as they were killing large numbers of native birds and small animals. The key message used was look what harm your cat is doing when it is wandering around at night. That seemed the right line to take, yes?

But in fact, it made very little difference and cat owners just kept letting their cats wander around at night. So – who you gonna call? Social researchers! After they conducted some research they found that cat owners were aware of the messages being circulated – but they just didn't care about native animals as much as they cared about their cats.

So a new message was devised, that aligned with this value – and cat owners were now told, don't let your cat out at night as it might be attacked, catch a terrible disease or even be killed.

And do you know what? People started keeping their cats in at night.

Getting just the right message and promoting a behaviour that is easily achievable is clearly very important.

But that's a simple case of behaviour change. And let's be very honest – there are also many forces of darkness out there trying to get people to behave in ways that might not be so good for them, or the environment, or society. This includes increasing the consumption of junk food, smoking or buying more stuff than a person needs or can afford.

In many cases you may be only one voice fighting against competing voices to get a certain behaviour change.

But keep your eye on your goal, start small and work on getting incremental changes of behaviour, measuring for impacts and modifying your strategies as you go.

Above all, stick with it and resist the temptation to lapse back into the comfort zone of information and education programs. For while these have been shown to be good at awareness raising, they are not so good at getting and maintaining behaviour change.[21]

'Behaviour extinction' refers to slow fade away of a particular behaviour when the stimulus or incentive for behaviour change is removed, such as when you notice the policeman who has been hiding behind the road sign all week is no longer there.[20]

Key summary points

- Behaviour change does not come easily and is hard to maintain.

- Changing a behaviour leads to changed attitudes much more effectively than a changed attitude leads to a changed behaviour.

- There are dozens of behavioural change theories and models and it is important to find the one that best works for your circumstances.

- Behavioural economics is based around using the learned logical patterns of human illogical behaviour.

Communicating controversies: The good, the bad and the ugly

'Benford's Law of Controversy states that passion is inversely proportional to the amount of real information available.'

– Gregory Benford, astrophysicist and sci-fi author

A long, long time ago, in a galaxy, not so very far away from this one, a scientist named Galileo Galilei was tried by the Roman Inquisition for stating that observations made through his telescope showed that the earth rotated around the sun, not the other way around. In 1633 he was found 'vehemently suspect of heresy' and was sentenced to indefinite imprisonment. He died still under house arrest, nearly 10 years later.

Thank goodness we've come a long way since then.

Well, unless of course you are an Italian seismologist. In 2009, six of them were jailed on charges of manslaughter for failing to predict an earthquake that hit the historic Italian town of L'Aquila, killing over 300 people.

Or unless you are a Taiwanese meteorologist. In the 1960s, four Taiwanese meteorologists were charged with 'dereliction of duty' for failing to predict a deadly landfall, following a typhoon that claimed 330 lives.[1]

Who'd have thought seismologists and day-to-day meteorologists were working in such highly sensitive and controversial areas?

But, every now and then – whether you choose it or not – there's a strong chance you are going to get drawn into a science controversy. It may be something that has to do with your own research, or researchers whose work you are communicating about, that is inherently controversial (hands up everyone working in evolution, coal seam gas, genetics, nanotechnology, biotechnology, artificial intelligence, quantum physics, anything to do with chemicals, climate science, animal research ... In fact, it might be a shorter list to ask hands up anyone who is working in a field that that they think is *not* sometimes controversial).

What is a controversy exactly?
As the US National Academies of Sciences, Engineering and Medicine politely put it in their report, *Communicating Science Effectively*, 'Although the public has a generally positive attitude toward science and scientists, specific contentious issues with a science component often become controversial.'[2]

I think we all know what that means in plain English: whether your science is controversial or not, it might well get sucked into a controversy. And that means controversies can exist beyond those working in those fields of GM foods, nanotechnologies, windmills, synthetic biology research … (hey, let's just say 'trigger science' and not have to repeat that list every time).

So it's worth considering what makes something controversial. I think we all know that it is something that upsets somebody else. But we need a bit more than that to work with. Sociologist Dorothy Nelkin has categorised several conflicts that underlie scientific controversies:

- benefits versus risks (such as company benefit versus consumer risk, for many GMOs)
- regulation versus freedoms (such as access to alcohol and tobacco or other drugs)
- science versus traditional values (such as occurs with alternative medicines)
- political priorities versus environmental values (such as nuclear power)
- efficiency versus equity (such as placing health facilities in big cities, far from many rural dwellers).[3]

The US National Academies has its own list of what underlies a science controversy:

- the fact that the science is inconclusive
- there are disagreements within the scientific community about the science
- there can be a disconnect between what science shows and what long-held common-sense perceptions indicate
- the sources of funding of the science implies a conflict of interest
- the science has been falsified
- the science offends moral, ethical or social values.

This last point can be particularly tricky, as different individuals may have differing ethical and moral principles, as can as different cultures and countries. And since we all know we are multicultural in the modern world, we need to understand that some people will not just disagree because they have a different ideology, but also because of a differing set of social values.

But trying to second-guess these doesn't always work either. For instance, if you were asked to guess if pig implants in human surgery were acceptable to Jewish or Islamic believers, what might you say? An Australian study looking at the use of animal parts in surgery found that Jewish leaders accepted the use of pig implants in human surgery, and Islamic patients were also permitted to use them (however, only in dire situations and only after all other options had been exhausted). But you can't extrapolate that across other beliefs, because Hindi leaders were not willing to accept body implants from a cow.[4]

We also know that different countries have different attitudes to different sciences and technologies, such as the global attitudes to genetically modified foods or climate change. In general, climate change is a major controversy in the US – and in Australia – but it is not in Europe. However, in Europe GM foods are a bigger issue than they are in the US or Australia.

According to the US National Academies report, science-related controversies often share three features:

- conflicts over the beliefs, values, and interests of individuals and organisations
- public perception of uncertainty either in the science or its implications
- the voices of organised interests and influential individuals are amplified in public, at the expense of statements on scientific evidence.[2]

Let's look at each of these, as they each provide unique challenges for science communication.

Conflicts over the beliefs, values, and interests

It comes as a shock to many scientists to discover that many concerns about science controversies are not actually about the science itself – which is why explaining the science does little to assuage any concerns. Most public controversies are all about how the science is going to be used, or who is going to benefit and who is going to get the risks, or whether it should be conducted in the first place.

We should know by now that any science or technology that challenges a person's individual beliefs or values is going to be contested. Those who oppose your science might be very well informed on it, and understand the science well, but fundamentally disagree with what is happening.

There are some who are opposed to GM foods who are quite scientifically informed, but it conflicts with their fundamental beliefs, such as all nature being sacred. In an ideal world, those who oppose a science or technology and those who support it could simply choose to use it or not. But we don't live in that world, do we?

Due to economic, social and political needs, we tend to find that those who support or oppose any science or technology go out of their way to convince others that they need to support or oppose it as well. And that means a battle ground of information and misinformation.

The general stereotype is that science-based bodies argue their case with science and facts and NGOs opposed to a particular science or technology argue theirs with emotions and misinformation – but that itself is a value judgment. If you spend time talking with NGOs who are characterised as being anti-science, you'll see they often strongly believe that they are pushing the truth and the science advocates are pushing misinformation.

The fact is they are both right. Some science agencies are more scrupulous than others when it comes to communicating the science and its impacts, and some less so. And some NGOs are more scrupulous than others when it comes to communicating their point of view.

Fighting for hearts and minds

It is just the way of the world that we are more likely to gain political victories if we have a large support base, so whatever the controversy it is useful to gain support, which means winning more hearts and minds than your opponents. Or it can simply mean making it look like you have won lots of hearts and minds. There are several NGOs in particular that have a tiny, tiny base of support – or can in fact be just one or two people – but work very hard to look like they have a large support base. With a busy website and an active social media presence it isn't that hard.

For some issues, such as embryonic stem cell research or evolution (in the US), religious beliefs are the dominant influencer.[5] In these cases people may feel that their religious views are being challenged, and therefore choose to reject the scientific findings.

There are many things we know about how people think that governs the support or adoption of different science controversies:

A zombie demonstrates how to win hearts and minds

- The theory of **motivated reasoning** says that people tend not to adopt explanations that conflict with their long-held views or values (such as those who treat nature as sacred opposing GM technologies).
- Our **cultural biases**, such as whether we prefer equality over authority, or individualism over community, can also determine whether we support a science or not, if we see it clashing with our cultural biases or not (infant vaccination is more aligned with individualism than community for instance).
- **Social identities** define us, and we tend not to adopt positions counter to those we believe are held by members of our group (which is why so many attitudes to climate change can be defined by support for political parties).

Fortunately, research has been able to identify several strategies that you can use to help mitigate the effects of competing beliefs, values and interests. Such strategies include framing and engagement.

You need to present information in a way that is consistent with people's values, as is outlined in Chapter 19, and engagement should be undertaken early, and scientists and interested parties should work together. The US National Academies report recommends:

> In general, for public engagement to achieve its potential, care must be taken
> to design a process that is attentive to the character of the issues at hand and
> that takes into account the strengths and weaknesses of individual thinking
> and group interactions.[2]

Manufacturing a controversy

It should come as no great surprise to know that some controversies are, in effect, manufactured. In his book *Creating Science Controversies*, David Harker says manufactured controversies are those that seem real from the public's point of view, but are actually not debated within the scientific world[6] – such as infant vaccination, GM foods and AIDS. Dissent occurs where there is overwhelming consensus by scientists on a position, but a lot of public concern or doubt.

Fake news and astroturfing

Artificially creating the perception of grassroots support is known as astroturfing and is an adjunct to fake news. It can be as simple as writing letters to the editor under fake names, or pretending you have a huge following through Twitter and Facebook campaigns, or more maliciously stealing identities and using them to push a line. You can even buy thousands of Twitter addresses to post your message out.

Twitter has admitted to the US Senate that they block almost half a million new accounts each day that they suspect are being generated through automation.[9] Clearly they miss a lot as there are companies that offer 20 000 retweets for as low as US$25.[9]

Early attempts at astroturfing by large PR companies working for tobacco groups included sending out hundreds of postcards to politicians, supposedly by concerned citizens, but in the digital age it has become more sophisticated and harder to tell the real from the artificial.

In 2006, a science journalist for *The Wall Street Journal*, Antonio Ragalado became suspicious of a YouTube video entitled 'Al Gore 's Penguin Army'. The video runs for about two minutes – so check it out if you have a moment – and it shows Al Gore's head superimposed on the body of Batman's enemy the Penguin, as he bores penguins and then hypnotises them to believe in his climate change messages.

The video is a little clumsy and looks like the work of any number of amateur animators around the world, working out of their bedroom or basement. However, this was actually a very slick production, designed to look amateur, and was produced by the Washington DC-based PR firm, the DCI Group – a company whose clients include ExxonMobil and General Motors.[10]

Ragalado became suspicious of the authenticity of the video when he found it was the first sponsored listing when he did a Google search for Al Gore. That meant someone was paying big bucks to promote it – and he then suspected, rightly, that someone had also paid big bucks to actually produce it.

The purpose of astroturfing, like much misinformation, is really just to throw doubt into the general mix. For doubt is very influential in science debates.

As Naomi Oreskes and Erik Conway point out in their ground-breaking book, *Merchants of Doubt: How a Handful of Scientists Obscured the Truth on Issues from Tobacco Smoke to Global Warming*, you don't need to prove a point when you are manufacturing fake news around a science controversy – you just need to throw in enough doubt that something is safe, or is having the impact that science says it is.[11]

And of course, it is very difficult for members of the public to know what is actually a scientific controversy within the scientific world and what is not. The public don't often read scientific papers or attend scientific conferences. Not even all scientists are across the breadth of science issues being discussed.

So what is a member of the public to think when they are told that nano-sized particles in sunscreens might be dangerous for them? They don't have ready access to the science discussions on this, and so it is plainly easier to avoid using the sunscreens, right? I mean, if someone says it's dangerous and a lot of the scientists give these long 'yes but', 'well but', 'perhaps but' and 'more research is needed' answers, then you know

Working with interest groups

Interest groups come in all shapes and sizes and are like the kid in the classroom who wants the teacher to pick his or her favourite game to play. They might tell the teacher that everyone else wants to play that game too, but don't want the teacher to actually ask anyone. They might tell the teacher that it is the safest game to play and won't cause the teacher any trouble. They might tell the teacher that if another game is chosen there is going to be a bit of an uprising amongst the students. And they may just tell the teacher that if they don't play the game that student wants to play, they are going to sulk and walk out of the classroom.

Interest groups tend to come in three different types, whether they are environmental NGOs, right to life groups, or fronts for industry:

1. Those who represent their members, such as patient's advocacy groups that hear what their members want and then go work for it. Or when that student has actually heard what game the students want to play and works to convince the teacher that is what is preferred.
2. Those who represent the interests of their members, such as disability advocates who might not have the capacity to consult with all members, but have a good general understanding of what is in their interests. Or when that kid knows that the teacher prefers to play a game the rest of the kids hate, and so tries to talk the teacher out of it in the interests of the other kids.
3. Those who really represent their own ideological interests. Some activist NGOs that have a constituent base that they report back to, or inform, but expect the membership base to follow their lead, rather than represent them or their interests. Or that kid who only likes one game, and doesn't care if nobody else likes it, and is prepared to go to extreme lengths to ensure that is the game that is chosen by the teacher.

Interest groups also tend to have three different ways of operating:

1. Those that are radical in public, but moderate in private. They are out there screaming at the TV cameras and threatening blockades and boycotts, and so on, but when you sit down with them to talk about things, they are quite reasonable to work with. They are fairly politically astute interest groups.
2. Those who are moderate in public and moderate in private. They make sensible comments to the media and then when you sit down with them, there are no surprises. These tend to be well established and used to working with different agencies and governments and so on.
3. Those that are radical in public and radical in private. They are out there screaming at the TV cameras and threatening blockades and boycotts, and when you sit down with them to talk, they are still screaming and threatening blockades and boycotts. These tend to be the hardest to work with.

that scientists don't agree on the safety, and you should err on the side of caution. That makes sense, doesn't it?

Playing the bullshit card

In scientific controversies it is not uncommon to find someone playing the misinformation card. And in the modern world that card can be harder and harder for people to identify. When you are working on a sensitive topic and people have strong emotional investments against the science you are working on, there are some common problematic cards that you might see dealt:

- the **threat to jobs card** that says this science and technology will take away people's jobs (such as automated robots)
- the **this is morally wrong card** that says the area of science should not be explored as it against too many moral principles (such as embryonic stem cell research)
- the **trust card** that seeks to undermine trust in the science, its funding, or the veracity of the findings (such happens with GM research a lot)
- the **doubt card** that says we just don't know enough about this to make a definite call on it, often holding up the 2% or 3% of dissenting scientists as evidence of doubt (such as the dangers of tobacco, or sugared drinks)
- The **bullshit card** (more politely known as **the misinformation card**), which is often played as a card of last resort, and says things that are just not true – but sometimes sound as if they might be true (again many statements about GMOs and climate change fall into this category).

The bullshit/misinformation card is played by all shapes and sizes of people and groups. Big companies with interests in coal and oil have been caught playing the card on climate change issues, small NGOs have been caught playing it on GMOs and nanotechnology scare stories, right to life groups have been caught playing it on embryonic stem cell issues.

Even science groups have been caught playing it, overpromising the benefits of a technology, or underplaying potential risks to get a technology accepted by an investor or government or the public. Remember how GM foods were going to feed the world, or how nanotechnology was going to completely revolutionise all fields of industry, or even how nuclear bombs could be used for peace by being used to dig harbours or even a new Panama Canal![12]

It's unfortunately true that misinformation is not limited to truculent children and interest groups who are trying to put their own spin on things. Science agencies are guilty of misinformation at many levels:

- over-hyping the impacts of some research
- misusing statistics
- stating conclusions beyond the data presented
- using causal language when not justified by the study design.[13]

Michael Blake of the University of Washington has said that misinformation, or bullshit as he happily calls it, can be more damaging than lies, as we often know it is incorrect, and start to think of facts as anything that can be adjusted until they match our chosen view of the world. This, he says, leads to all political disputes – and in this case science disputes – being about moral world views rather than about facts. These type of conflicts have 'been the source of our most violent and intractable conflicts'.[14]

Unless of course you happen to know that there are campaigns being waged against all things nano that question the safety of everything that has any nano-sized particle in it. Which have led some people to stop using sunscreens altogether for fear of the nanoparticles in them, and put themselves at a real risk of developing melanoma.[7]

That doesn't make a lot of sense, does it?

If you want to see just how easy it is to manufacture a controversy around science, check out the several YouTube clips on the dangers of di-hydrogen monoxide. It is clearly a fairly dangerous chemical, as it is a component of acid rain, it can cause severe burns in its gas state, it can be toxic in even small quantities and most people interviewed about it believe it should be prohibited from schools, regulated and banned! The only problem is that di-hydrogen monoxide is H_2O – water!

Some of the tactics you might more commonly see in manufacturing controversies include:

- taking quotations out of context to make scientists sound like they question consensus findings
- citing small disagreements among researchers to argue that the entire field does not support the major conclusions
- highlighting small errors by individuals to argue many of the findings are false
- disputing accepted scientific evidence, rather than doing any original research
- promoting 'alternate beliefs' that often lack any scientific basis.[8]

Dealing with uncertainty

Many scientific controversies are characterised by the science around a topic being seen to be uncertain or unclear, which is exacerbated when there is debate within the scientific community itself. Some such issues include:

- causes and impacts of obesity
- health impacts of e-cigarette vaping
- cellular or mobile phones
- gene therapies.

And it only takes a tiny bit of uncertainty to be planted and grow. Added to this of course is the nature of science itself, where certainty is rare. And point me at a researcher who has not ended a scientific study with the phrase 'but more research is needed'. And any inherent uncertainty is often exploited during controversies.

Another way that uncertainty is magnified during controversies is through the media's habit of needing to show both sides of an argument and presenting a dissenting voice as often having equal weight to the majority of scientific consensus. For an interest group whose sole aim is to sow uncertainty, that dissenting voice need not even be a relevant one. Many scientists who speak out against human-induced climate change are from fields that only have small relevance to climate studies.

The best way to address uncertainty, without arming those who would use it for mischief, is through a framing that acknowledges that science cannot always guarantee 100% certainty, but that there is general scientific consensus on a particular topic.

Researchers have sat people down and shown them two similar statements on climate change – one indicating the high consensus among scientists, and one giving a more general statement. They found that those who learned there was a high scientific consensus on climate change causes were both more likely to believe it, and also more likely to support policies aimed at its mitigation.[15]

So what can you do about it?

There is a lot we know about science controversies and who might drive them and so on, but how does that help you deal with one? The general advice is, do whatever you can. And keep doing it, even if you feel it is getting nowhere. As climate change has shown, hundreds if not thousands of scientists have been working for many years to try and

What to do with what you now know

There is a good chance that some time in your life you are going to be caught up in a science controversy – even if it is as simple as defending the science behind infant vaccination or climate change at a party. And those are good forums for practicing communication strategies.

A good checklist to work with is:

- acknowledge there are different world views that drive different perspectives, not different evidence
- maintain a coherent story
- reiterate the nature and findings of the science
- don't get drawn into a fight or argument – rather simply state you disagree based on evidence
- correct misinformation politely without belittling the other person, though you can question the reliability of dodgy sources of information
- accept that you are not going to get anywhere with a person who is a 'rusted-on' believer in an anti-science position and find a way to agree to disagree, but leave them with something to consider
- normalise scientific uncertainty and dissent, and acknowledge that it often exists in science and is not a sign of any concern
- don't get drawn into being a moral or ethical commentator if you have no authority to do so.

When getting involved in science controversies a draw can be a good outcome, and too often both sides walk away thinking they have won because they got to repeat their points so often – but let's be frank, that is measuring the wrong thing. You are not going to change a person's mind about a science controversy in one conversation – but you might get them to question what they have been told and accept your position too, if you can maintain reasonableness, respect and back up your arguments with good science.

make the science more relevant in the face of political and media controversy, with only limited impact.

But what might the situation be like if they had not spoken up?

Many controversies eventually reach closure though several factors: closure through interest fading away, closure through consensus or negotiation, closure through force or regulation, or even closure though sound arguments winning out.[16] But the fact is that for many controversies, the best result you can expect is a digging of the trenches of those for and against and sitting in them sniping at each other. But even then, you should keep talking.

And I know I have said that data rarely stands up well against emotion. Well sometimes data wins. For instance, if you can find statistics or data on public attitudes towards the controversial science, use that to respond to interest group claims about what the public wants. I remember talking at a forum on genetically modified foods where an anti-GM advocate stood up and said, 'No one in Australia wants this technology. No one in Australia thinks this technology is safe. No one in Australia trusts the science behind it.'

And then I stood up, with the latest findings of a public attitude poll we had done and said, 'Well in fact that's not what the data says. An independent survey shows that more people support GM foods than oppose them, and most trust the regulators.'

It can be a big victory for data!

Key summary points

- Science controversies are often underpinned by clashes of values.

- Controversies are exacerbated by uncertainty.

- Some controversies are manufactured by interest groups, who work hard to create uncertainty.

- There are some recommended strategies for communicating during scientific controversies, based around creating trust, acknowledging uncertainty and engaging with audiences.

22

Debunking bunkum

'I feel I change my mind all the time … And if you don't contradict yourself on a regular basis, then you're not thinking.'

– Malcolm Gladwell, journalist and author

This really should have been Chapter 13 – looking at superstitions and pseudoscience and the rather peculiar anti-science beliefs some people have, and what you can do about them. But it's 22. Unless we all just pretend it is 13. I mean I think we've learned by now that most of us never let the facts get in the way of a preferred finding.

So, triskaidekaphobia – a fear of the number 13 – is surprisingly prevalent across Western societies. At least 10% of people have some concern or worry about the number 13, ranging from those who regard it with caution like a new in-law, to those with a debilitating fear of it.[1] This fear appears to be culturally acquired, because in some cultures, such as China and Ancient Egypt, it was considered a lucky number.

Historically the bad press that number 13 gets has been attributed to the Last Supper, where Judas, who betrayed Jesus to the Romans, was said to the be the 13th person at the table. Another possibility is because it is just different to the in-group of 12s – like 12 months of the year, and 12 symbols of the zodiac and 12 hours on a clock and so on.

Many hotels, you might have noticed, do not have a 13th floor, as guests do not like staying on that floor. There is even data that more people avoid going to work on Friday the 13th, and productivity drops on that day. It is estimated that the United States loses about $900 000 000 in productivity every year because of Friday the 13th, because some people are so superstitious about it that they wouldn't even get out of bed.[1] That is a double whammy – fear of 13 and fear of Friday together – known as paraskevidekatriaphobia.

True fact!

Other common superstitions include it being bad luck to walk under a ladder, breaking a mirror or having a black cat cross your path, and those that bring good luck include finding and picking up a penny, and the number seven. These can be very widespread, with 33% of those living in the US believing in the lucky penny for example[2] (see Fig. 22.1).

Looking a little wider, at paranormal beliefs, a 2005 poll in Canada found that 47% of Canadians believed in ghosts, and a 2014 poll in the US found that about the same

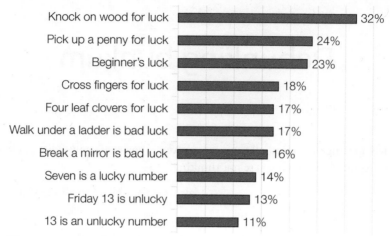

Fig. 22.1. The percentage of people who believe in good and bad luck[4]

number believed in ghosts there.[3] So clearly no issues of ghosts being stopped at the border.

And there is actually evidence that some superstitions can work in your favour. A team of psychologists at the University of Cologne in Germany reported that students who relied on good luck charms did better in tests than those who did not. The researchers had given participants a golf ball, of all things, and those who were told that the ball was lucky tended to outperform those who were not told that.[5]

That's all very interesting, but what's the problem?

Well, the problem is that the same irrational but strongly held beliefs that underpin superstitions can drive anti-science beliefs and make them just as difficult to change in people.

Really? To quote a science communicator colleague of mine, 'Absoloodle!'[6]

A study undertaken in 2012 to understand better how pseudoscience rationales can affect the way we judge things had education, science and nursing students in Canada and the UK look at one of three versions of a news article. The story was about a ghostly apparition of a murdered girl appearing in a photograph of an old school building.

The first version of the story simply stated it was a paranormal event.

The second version referred to the author of the article as a PhD science correspondent, and gave an explanation for the image of the girl using pseudoscience.

The third version also referred to the author as a PhD science correspondent, but gave a scientific rebuttal of the phenomenon.

And the result? First, the researchers found that a large number of students arrived at university with significant beliefs in magic and that studying science and empirical evidence did little to diminish them. The second surprising find was that the use of pseudoscientific explanations or scientific explanations did not influence their views to any large degree.

The researchers concluded:

> This supports the view that the criticality and scientific rationale engendered during an education process may be insufficient to influence a pre-existing personal belief based on personal experience.[3]

So if you have pre-existing anti-science paranormal beliefs it will lead to an increased likelihood of finding paranormal reports scientific, believable and credible. Something astronomer and pioneering science communicator Carl Sagan referred to as our demon-haunted world!

Looking at several well-established scientific theories that attract anti-science beliefs, such as the theories of relativity and evolution, as well as climate change, Swedish philosopher and sceptic, Sven Ove Hansson has said that science denial is a form of pseudoscience. He has also said that such science denial is characterised from other forms of pseudoscience through persistent fabrication of fake controversies, the extraordinary male dominance among its activists, and its strong connection with various forms of right-wing politics.[7]

When pseudoscience is dangerous

Believing in ghosts mightn't really be such a problem for society but believing in fringe medical beliefs might be. For example, US health blogger Vani Hari, known to her zillions of followers as the Food Babe, regularly claims to research and reveal problems with food. She does not reveal as frequently that she receives a lot of sponsorship from 'natural' food companies.[8] One of her research findings about the impacts of using a microwave oven was:

> Microwaved water produced a similar physical structure to when the words 'Satan' and 'Hitler' were repeatedly exposed to the water.[9]

Another blogger, Yvette d'Entremont, known by her own zillions of fans as the Science Babe, has described the Food Babe as 'utterly full of shit'.[10]

Or Australian wellness blogger Belle Gibson, who advocated a range of wholefood recipes and alternative therapies. She held up her own fight against cancer as proof of their effectiveness – a cancer she later admitted that she'd entirely made up.

Or Jessica Ainscough, known by the blog name the Wellness Warrior. She promoted natural healing, including coffee enemas, to treat her own sarcoma. She also promoted a treatment that blogging surgeon Dr David Gorksi described as the 'cancer quackery of cancer quackeries' – the Gerson therapy (which involves a plant-based diet, raw juices, coffee enemas and natural supplements).[11] Tragically, they didn't work for her and she died in 2015 of sarcoma.

All these bloggers tend to be very photogenic, gorgeous even, and emanate the aura of celebrity – and the associated sponsorships, book deals and television appearances that go with that, which we know increases credibility and trust. And celebrities are people that many of us would like to have as our friends, which increases our trust in them too.

Professor Christian Behrenbruch, of the Royal Melbourne Institute of Technology, says he dedicates at least three hours a day to dispelling pseudoscience. Whenever there is money involved, science gets thrown out the window, he says.

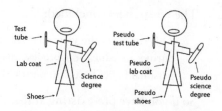

He also says that when people have made a financial or other commitment to a dodgy science it is very difficult to convince them they are backing something dodgy. And for people with

How to tell a scientist from a pseudoscientist. Can you spot the difference?

serious health issues, as their condition degrades and there are fewer and fewer treatment options, they will invest in hope – and that, he says, undermines science every time.[12]

Debunking misinformation

Many science communicators and scientists feel this great and overwhelming urge to correct any information that is inconsistent with the weight of scientific evidence. However, such attempts can actually backfire and end up reinforcing false beliefs, especially among the more educated.[13]

Trying to correct any false beliefs can be very difficult when incorrect information is consistent with how people think about something, as has been demonstrated in the climate change debate so often. Research has even found that when you challenge people about their beliefs, they may just dismiss the credibility of the messenger who is trying to correct their beliefs.[14]

Even people who are followers of conspiracy theories find that these enable them to make some sense of the world, so challenging those conspiracies leaves them without that comforting sense of feeling how the world works. You can't take such things away from people without giving them something more meaningful in return.

Researchers John Cook and Stephan Lewandowsky have produced a great guide for those interested in debunking, called *The Debunking Handbook,* which you can find online. It looks at things that will more likely work in trying to debunk bunkum and which things are least likely to work. They state that any effective debunking requires three major elements:

1. The refutation must focus on core facts rather than the myth to avoid the misinformation becoming more familiar.
2. Any mention of a myth should be preceded by explicit warnings to notify the reader that the upcoming information is false.
3. The refutation should include an alternative explanation that accounts for important qualities in the original misinformation.

Attempting to get people to swap a fringe belief with high appeal for an evidence-based one with low appeal would be about as successful as asking a scientist to swap their evidence-based beliefs for fringe ones.

You might think that repeating the myth is okay, but in fact research shows that it just concretes it in people's minds.

Really? Absoloodle!

One experiment, quoted in the handbook, involved exposing people to misinformation about a fictitious warehouse fire, then giving a correction clarifying the parts of the story that were incorrect. The researchers found that despite remembering and accepting the correction, people still hung onto the misinformation when answering questions about the story.[14]

This is known as the continued influence effect and explains why you don't give air time to the incorrect story or argument.

One of *The Debunking Handbook* authors, Dr Stephan Lewandowsky, gives another example on the blog *Skeptical Science*: 'It is false to claim that Listerine alleviates cold symptoms.' This statement unavoidably strengthens the link between the two concepts, namely Listerine and alleviating colds, he says, even though the statement seeks to correct that myth.[15]

Other research has found that providing overly complex information to try to correct a more simply worded myth is likely to be ineffective. It becomes a battle that the simplest message will win out – which there are plenty of examples of in popular politics.

Prebunking

The US National Academies report *Communicating Science Effectively* states that one approach to avoiding the risks of debunking is to focus on messengers instead of their messages, as distrust of a communication can help combat the effects of misinformation.[16] This can involve 'prebunking'.

Prebunking can be thought of as inoculating audiences against the intentional efforts of individuals or organisations to mislead the public. There is a neat symmetry in that you can inoculate an audience against misinformation on vaccinations, for instance. This is done by warning people that they may be exposed to misleading information and giving some explanation as to why it is being promoted to them.[14]

When there are competing stories floating around in public, we have already seen that the simplest story, and the one that most aligns with current beliefs, is the one that is more likely to get picked up, but large public health campaigns have shown that it also helps to blitz the mediascape. Have a lot of exposure, early on, through a variety of channels, and that can help ensure your perspective is heard and considered.

Getting opinion leaders to support you can also be influential – but make sure you pick the right ones that work for your audience, not just for you.

But we need to be aware too that there are some conflicts that arise in attempting to debunk misinformation and replace it with science-based information:

- How do you raise trust in science, but keep some scepticism in science too?
- How do you tell people what they should believe but expect them to learn things for themselves?

What to do with what you now know

People hold onto alternative beliefs because they affirm their world view, and if you debunk their beliefs as bunkum, you will need to make sure you are offering something more acceptable to them as a replacement. For instance, if you are going to try and tell someone that their fears about GM foods are unfounded, because enough tests show they are safe, you are also going to need to address their concern about multinational control of the food chain, or the need for many more years of testing, among other possible concerns.

If the underlying belief is that science and technology is working against nature, you will need to convince the person that GM foods are in the best interest of nature. This might be done through telling them that GM is one of the only ways that crops can adapt to changing climates or droughts.

Or if you were to tell someone that infant vaccinations are not harmful, you will not only have to find a way to frame that as not impacting their free choice, but you will need to also address fears over government control or the profit-seeking motives of big pharma.

You need to unpick why people have an anti-science position, and offer them an alternative science-based belief that is at least equally acceptable to them. Less acceptable won't do. And arguing the science is only a part of the job (and labelling anything as anti-science will only work on those who respect and trust the science).

And yes, that is a lot easier said than done, and is generally never achieved after a single engagement. But that is not reason not to try, yes?

Teaching the scientific method?

Many scientists advocate a stronger teaching of the scientific method as a part of all science education, so that young people understand better how evidence is accumulated and how scientific conclusions are made. But, as was pointed out in the ghost study above, there is evidence that even this might not be enough to turn around anti-science beliefs that are pre-existing and strongly held.

Margaret Defeyter, from Northumbria University in the United Kingdom, says if you want to dispel myths in people's minds you need to improve your ability to communicate, using more creative approaches such as hands-on activities that encourage self-directed learning. She says, 'Rather than just trying to stamp out misunderstandings, we need to offer people something else to believe in.'[12]

Absoloodle!

Key summary points

* The same irrational but strongly-held beliefs that underpin superstitions can drive anti-science beliefs, and make them just as difficult to change in people.

* Many online celebrities have used their celebrity status to promote pseudoscience.

* Debunking pseudoscience can actually strengthen beliefs in it, if not done well.

SCIENCE COMMUNICATION ISSUES

23

Fantastic ethics and where to find them

'With great power comes great responsibility.'

– Stan Lee, *Spiderman*

So with all you've learned so far, and all the science communication super powers that you've acquired, we now need to consider just how to use those powers for the forces of good and not get tempted over to the forces which are not always so good. For while much of the best practice of science communication is based on the research that underlies social psychology and behavioural economics and social psychology and so on, there are also large corporations, political parties, governments and think tanks who are using the same research to push their own agendas. And some of them have budgets way in excess of what you'll ever have, or what any university research department will ever have, to spend in this space.

Erik Conway and Naomi Oreskes outlined some of the sneaky tactics used by tobacco companies and climate change critics in their book *Merchants of Doubt*.[1] Though we should not forget that the subtitle of the book was: *How a Handful of Scientists Obscured the Truth on Issues from Tobacco Smoke to Global Warming*. Yes, it was scientists who were skewing data in the interests of sowing some doubt and obscuring the truth. It happens.

And there is some awesome data out there on what influences us and what can make us behave and think in different ways, but we'll never see it as much of it is 'black data'. You are probably familiar with grey data – that which is gained from outside traditional scientific publications, such as annual reports and working papers – well black data is unseen. It is generally commercial in confidence and undertaken for large corporations or governments, and is not shared.

You think companies like Nike and Coca-Cola and Apple don't have reams of data on how to influence teenagers to buy their stuff? You think beauty product manufacturers don't know the insecurity they are seeding in people? You think Cambridge Analytica's manipulative misuse of Facebook data was a one off?

Sorry, but the forces of darkness really do exist, and they are working very hard to manipulate every one of us for their own benefits. The list of companies and institutions that have been uncovered as screwing over the consumers seem to grow each month:

banks, financial planners, insurance companies, superannuation companies, supermarkets, social media firms, telecommunications companies. Just add to the list whoever has been in the media or before parliament recently trying to explain and justify their behaviour.

And on that list are some scientists and social scientists and science communicators. And the tragedy is that some have slipped over to the dark side without even really knowing it.

Fighting the good fight in times of uncertainty

The post-truth, post-trust, post-expert world we are living and working in makes the question of ethics just that little bit more tricky. In times of divisiveness and clear manipulation it can be very tempting to use the same tactics yourself. But in the cause of good, yes?

Well, yes, until you aren't any longer.

The road to the dark side is taken in very tiny steps. You can start out with good intent and not even notice when you are over-hyping the data or trying to over-spin the good side of a story to combat the negative stories that exist.

Here's an example. The Fourth Assessment of the Intergovernmental Panel on Climate Change oversimplified data on possible sea level rises from disintegration of the Greenland and West Antarctic ice sheets to remove uncertainty. However, this resulted in projected sea level rises that were seen as misleading.[2] But the original simplification was actually done in order to deny the merchants of doubt that uncertainty, which they would certainly have manipulated, for it was believed that providing statements on uncertainty would have been playing into their hands. The end result though was that the report was seen as misleading rather than uncertain. So be warned – damned if you do and damned if you don't.

Framing ethically

We have already looked at framing (Chapter 12) and narratives (Chapter 7) and how they can help influence your audience by providing your story in a context that your audience is more likely to accept. But there is a very fine line between doing this so that the audience gets the message and doing it so the audience gets only some of the message and not the bits you don't want them to get.

Ask yourself what you might choose to do if you were running a public health campaign aimed at persuading teenagers to avoid smoking or binge drinking, if you knew that it was more effective to use emotional appeals to social norms that selectively presented information, rather than gave all the facts?[3]

According to science communication researchers Matthew Nisbet and Dietram Scheufele, science messages can be framed ethically when the framing is used to motivate greater interest and concern about an issue, expanding it beyond polarised and gridlocked views, or providing more context for dialogue. But when it is really just a strategy to promote the selling of science to the public it is less ethical.[4]

According to the US National Academies report, *Communicating Science Effectively,* any decision to communicate science always involves an ethical component. It states, 'Choices about what scientific evidence to communicate and when, how, and to whom are a reflection of values.'[3]

This is more so for science findings that impact policy or are seen as contentious, where the temptation to push one side of the story for a particular outcome is very strong. You see it a lot in areas like biotechnology, where science-based arguments go out of their way to deny other arguments. For instance, why were anti-biotech activist groups excluded from Canadian agri-food policymaking?[5] Was it that their data was shonky or was it that their input would have detracted from a sought-after goal?

And how is it that synthetic biology is promoted as being something really new and exciting, until people start talking about a need for new regulations or new causes of concerns, then it is suddenly framed as being just the same old same?

These are important questions to ask. After all, what is the real purpose of your communicating science to the public? Is it being done to inform and empower, or is it being done to persuade and influence people?[6]

Matthew Nisbet of Northeastern University in Boston recommends four guiding principles to determine the ethics of using framing techniques within science policy:

1. Frames should be used to emphasise common ground and promote dialogue, not to manage information.
2. Frames should clearly communicate the underlying values guiding a policy choice, not just suggest science information compels a decision.
3. Frames need to remain accurate and not distort or exaggerate the meaning.
4. Frames should not be used to negatively typecast any social group or political leaders for deliberately electoral gains.[7]

Speaking out on the big unspoken issues
And talking of things we need to talk about more, there are lots of issues around science itself that are rarely spoken out about. Perhaps for fear they will ultimately do more harm than good. For instance:

*   Clinical trials of new drugs tend to show poor translation rates from animals to humans.[8]
*   Too many scientists appear to be fudging data to get the statistical findings they need.[9]
*   More than 70% of researchers have tried and failed to reproduce another scientist's experiments, and more than half have failed to reproduce their own experiments.[10]
*   12% of medical research papers and 82% of humanities papers are never cited by other researchers.[11]
*   The greater the financial and other interests and prejudices in a scientific field, the less likely the research findings are to be true.[9]

- Claimed research findings may often be measures of the prevailing biases.[9]
- The chase for publication leads to conclusions that are not always warranted by the data.[12]
- The chase for media coverage leads to media releases that are not always warranted by the data.[13]

You know there is some problem when established journals like *The Lancet* come out with statements like, 'Much of the scientific literature, perhaps half, may simply be untrue'.[14] To prove this point Harvard biologist John Bohannon created a make-believe cancer research paper and submitted it to 304 journals. Some allegedly peer reviewed. It was accepted in 157 of them.[15]

And that naturally brings us to the growth in dodgy or predatory journals that allow you to publish dodgy research more easily without proper checks. To many members of the public or the media they still seem to be valid scientific publications. A multi-authored research paper on predatory journals published in the *Journal of Korean Medical Sciences* said:

> The ease of launching online journals in the Internet age creates short-cuts for illegitimate publishers and stand-alone journals, pursuing financial gain at the expense of the quality and validity of the publications. Papers in such journals may suffer from plagiarism and other forms of research misconduct that pass unnoticed due to the editorial negligence and lack of readers' attention.[16]

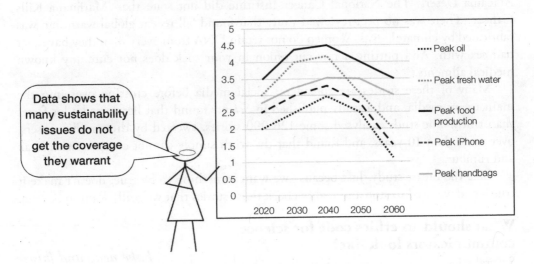

Playing media popularity

Another conundrum for science communicators is the pressure to play for media results, which can result in dumbing down stories to the point that they are inaccurate. Or worse, perpetuating stories that are both false and potentially harmful.

What choices would you make as a science communicator if you had to decide between a story around the hashtag #DoesItFart,[17] knowing that a database of animals that fart will attract attention, or a complex story on the impact of stress on epigenetics?

A study conducted by *The British Medical Journal* found that university press offices were a major source of overhype, with over one-third of press releases containing either exaggerated claims of causation, unwarranted implications about animal studies for people, or unfounded health advice.[13]

For instance, things that in the media have been stated to both cause and to cure cancer at different times include aspirin, cheese, grapefruit, mammograms, milk, pizza, pregnancy – and of course sex.[18] To address this, the UK Science Media Centre has implemented a labelling system for medical research press releases that states whether it has been peer reviewed, the type of study conducted, and whether it was conducted in animals or humans.[19]

There are a lot of great science stories that get run all over the globe, that later turn out to be entirely fictitious, yet science journalists, science communicators and scientists regularly refer to them. One example is the widely-quoted German research study, published in the esteemed *New England Journal of Medicine* that found that if men looked at large breasts for 10 minutes a day it would lead to health benefits including living longer.

Sorry fellas. Not true. The booby prize! A hoax.[20]

Likewise, there was no two-head alligator spotted in Florida. No alien found in the Atacama Desert. The National Cancer Institute did not state that 'Marijuana Kills Cancer'. There was no peer-reviewed study that found 'all recent global warming' was fabricated by climatologists. Women do not retain DNA from every man they have ever had sex with. And putting a raw cut onion in your sock does not cure any known medical ailments.[21]

Many of these stories are spread on social media before crossing over into the mainstream media, and a study published in *Science* found that lies spread a lot faster than truth. The study analysed some 126 000 stories, tweeted by three million users, over more than 10 years, and found that the truth simply cannot compete with hoax and rumour.

Like the breast study. Just because we want something to be true, doesn't make it true – and we need to stop and reality check those stories that we really want to be true.

What should an ethics code for science communicators look like?

Several scientists and science communicators have argued for a code of science communication ethics,[30] but others have pointed out problems with this. Joan Leach, of the Australian National University, and Fabien Medvecky, of the University of Otago in New Zealand, have argued rather for a set of principles rather than a code of ethics.

Fake news and false rumours reach more people faster and penetrate deeper into social networks than accurate stories do.[22]

Fact checking

There are some great fact checking sites out there that include:
- Snopes.com
- FactCheck.org
- Hoax-Slayer.net
- TruthorFiction.com
- SciCheck
- Climate Feedback
- HealthNewsReview
- Kill or Cure
- ThoughtCo
- Discovery channel's *Mythbusters*
- Rational Wiki
- QuackWatch
- International Fact-Checking Network.

We need to have a very serious talk about gender

We could do a whole chapter on gender issues in science and science communication, but instead we are going to rely on the super powers of metaphor and talk about the super hero Black Widow. You know the character played by Scarlett Johansson in the Marvel movies.

Well, one of the best scenes in the movie *Avengers: Age of Ultron* is when the Avengers' Quinjet swoops down low over the streets and Black Widow drops out of the bottom of the aircraft on her motorbike and hits the road and zooms off to fight evil.

Yeah! Kick-arse awesome!

Except when the toy companies Lego and Hasbro released their tie-in toys, suddenly it was Captain America who was dropping from the jet and riding the motorbike. What the … ?

The companies copped quite a serve in the media and online for replacing a female action figure with a male one, and their excuse was that female action figures don't sell as well as male ones. Which is a pretty lame excuse really – because how do you make change if you are not part of the change?

I really think Black Widow needs to knock on their door and say, 'We need to have a very serious talk!'

Or take the European Union's clumsy attempt to make a video promoting more women to take up careers in science with their 'Science: it's a girl thing' video, that put more emphasis on cosmetics and fashion than science. It generated such outrage it was gone in a few days.

There is data as well as anecdotes. Several studies have found evidence for significant bias against female scientists. Compared with their male counterparts, they:
- receive grants less often and receive smaller grant allocations[23]

- receive fewer citations[24]
- receive fewer scientific awards[25]
- are less likely to get promoted
- are less likely to present their research at conferences
- are less likely to publish their work or collaborate internationally
- are under-represented among inventors
- are less likely to hold academic leadership positions
- receive faculty recommendation letters that are less praiseworthy.[26]

It has even been stated that taken together, the data suggest a pervasive culture of negative bias – whether conscious or unconscious – against women in academia.[27]

So the ethics of gender are something you need to continually consider. The same as race and disability and other forms of diversity.

Ask yourself if you have really tried to get balance in representation in stories and activities you are doing. Here is a bit of a checklist:

- Do you have quotes from both men and women that are equal in the worth of the quote, and from equality in work roles?
- Are you supporting gender stereotypes or helping to break them down?
- Are you using gender-fair language, such as avoiding 'he' as a generic pronoun and words like chairman, businessman, mailman, etc.[28]
- Do your images promote equality of roles?
- Do you have a mission statement on gender equity and diversity to refer to?[29]

After all, how do we ever get change if you are not going to be a part of the change?

Because if not, you should know the day is definitely going to come when Black Widow knocks on your door and says, 'We need to have a very serious talk about gender!'

They state that many of the driving ideas behind science communication give it a sense of moral righteousness. And of all the roles that science communication adopts, the role of knowledge broker can be the most ethically fraught, as 'knowledge brokers direct attention to specific bodies of knowledge and away from others'.[31]

They state that as science communication is the child of so many parent disciplines – including communication, marketing, social science, journalism and so on – that the ethics guidelines that exist in these areas cannot be easily reconciled with each other. For instance, the CUDOS model (communalism, universalism, disinterestedness and organised scepticism) is a good guide for the conduct of science, but doesn't always provide practical value for many of the questions that science communicators need to consider. For instance, how does the model apply to issues of timing or hype? The same problem arises in trying to adapt journalistic or communication ethics guidelines.

Using social science to influence Congress to go to war!

So you want to know how this is done? Let's look at one of the few incidents that have come to light. Way, way back in 1990, Iraqi dictator Saddam Hussein invaded his oil rich neighbour Kuwait. This was a problem because although he was clearly an evil tyrant he was considered by the US as a good evil tyrant because he was a political ally. Sure, he had killed about 150 000 Iranians and about 13 000 of his own citizens – but you need to look after your friends, right? Well, until your other friends complain about it. Or access to oil is at stake!

Then you are no longer a friend. You are suddenly a hostile adversary.

John Stauber and Sheldon Rampton, authors of the expose of the PR industry, *Toxic Sludge is Good for You,* summarised the dilemma for the US Government as:

> *Bush would need to convince Americans that former ally Saddam Hussein now embodied evil, and that the oil fiefdom of Kuwait was a struggling young democracy. How could the Bush Administration build US support for 'liberating' a country so fundamentally opposed to democratic values? How could the war appear noble and necessary rather than a crass grab to save cheap oil?*[34]

Enter the major US PR firms with a campaign that made Operation Desert Storm look clumsy. As many as 20 PR, law and lobby firms were employed by the Kuwaiti Government to mobilise US opinion against Iraq. And the firm that coordinated it all was Hill and Knowlton – at the time the world's largest PR firm.

It was at the time the largest foreign-funded campaign ever aimed at manipulating American public opinion.

The Kuwaiti Government worked through a front group, Citizens for a Free Kuwait, funnelling almost US$12 million. Of this, almost US$11 million went to Hill and Knowlton. Also of interest, the man who was running Hill and Knowlton's Washington office was Craig Fuller, a close friend of President Bush and his former Chief of Staff and political advisor.[34]

Using advanced polling and focus group testing the campaign was able to test the types of messages most likely to resonate with the American public, the media and with Congress, and pushed out those messages in print, video and radio.

But the big goal was getting congress to press the button to go to war. And that's when things got really 'interesting'. Getting the general public and the media on your side is a big help, but imagine knowing just how Congress will act because you have not only lobbied congressmen and women incessantly, but you have data on the way that people with the same demographic and psychographic profiles of congressmen and women think? Imagine having enough good data that you could, in effect, create a data match of Congress, and then discover what types of Iraqi atrocities might make them hit that button?

Shooting civilians? Nope. That happens in all wars. Using chemical weapons? Hell – Saddam Hussein had been gassing Kurds in Iraq for years anyway. What about atrocities on babies? – Well now you you're talking. Give me that button!

The key moment was in October 1990 when the Congressional Human Rights Caucus held a hearing on Capitol Hill into human rights abuses in Kuwait. And the star witness was a 15-year-old Kuwaiti girl, known only by her first name, Nayirah, to protect her family from

Iraqi reprisals. Breaking down in tears she told how she had been volunteering in a hospital in Kuwait City and had seen Iraqi soldiers come into the hospital with guns, and go into the room where '... babies were in incubators. They took the babies out of the incubators, took the incubators, and left the babies on the cold floor to die.'[34]

So Congress went to war. Though only by five votes, which was a narrow thing.

It was only later revealed, however, that Nayirah was in fact not only a member of the Kuwaiti Royal Family and was most likely not even in Kuwait at the time she claimed, but her father was actually Kuwait's Ambassador to the US.

In trying to defend the fact that she had been coached in her testimony by Hill and Knowlton, her father, Ambassador Saud Nasir al-Sabah, revealed a little too much when he stated, 'If I wanted to lie, or if we wanted to lie, if we wanted to exaggerate, I wouldn't use my daughter to do so. I could easily buy other people to do it.'

The researchers conclude that science communication can't easily call on any existing 'off-the shelf' ethical guidelines because fundamentally, science communication is neither science, nor journalism, nor straightforward communication.

Fabien Medvecky has argued for adopting principles similar to those used in bioethics, as they can be situationally dependent and apply to both research and practice.[32] These principles are commonly seen as:

- respect for all persons
- do no harm
- beneficence
- justice.[33]

And yet there are still some issues they might not address well, such as the divergent pulls of wanting to do the right thing, as a communicator, but being beholden to agendas of the agencies we might work for (not that this is particular to science communication).

What to do with what you now know

If only this were as simply as being able to reset your moral compass to the setting of 'be ethical' and let it guide you. But ethical issues are often complex and nuanced and you can easily end up in an 'ethically difficult place' in tiny little steps that are not discernible – or through a process of group-think that leads you to make dumb decisions that seem okay at the time.

One technique I recommend is that discussed in the zombie movie *World War Z* (strange as that may sound). Whenever you have an accord on any issue appoint someone Devil's Advocate, to ask all the difficult – and ethical – questions. If you can answer them all, that is fine, but odds on there are going to be some things that are not so easy to answer – like, are you sure this is not going to disadvantage any one group over another? Or, will everyone have equal access to this technology?

It can be a bit uncomfortable – but it will really sharpen your chances of surviving the zombies! (Yes, that's clearly another metaphor!)

So what should we do?

I think we should know by now that effective science communication is based on trust and in a post-trust social environment it is crucial to maintain and protect trust. So any attempts at manipulating the public through narratives and framing techniques for example, are likely to end up in tears in the long run if they damage your trust.

It is no longer sustainable to live on the social credit of science, as it is slowly diminishing. This is due to a variety of factors, including the change from science being primarily done for social benefit to be done primarily for institutional or corporate benefit, and the pace of scientific developments outstripping the comfort of many in the community.

We don't yet have the answers as to what ethics guidelines and principles for science communication should be, but the fact there are serious discussions around them is something that can be made public to help maintain trust in the profession.

Key summary points

- There are many ethical issues to consider in science communication, often based around how to best inform someone using framing or narratives, without manipulating them.

- We should be clear on the purpose of science communication and not confuse informing and empowering with persuading and influencing.

- The push for publication and the push for media coverage has led to many shoddy research results being published.

- Adopting ethics frameworks from other disciplines is problematic as science communication encompasses so many different disciplines.

- A better way to consider ethics might be to use broader principles rather than specific guidelines.

Why we don't actually know what we think we know

'The more I learn, the more I realize how much I don't know.'

– Albert Einstein, icon

Here's a very, very big question. What if everything we have learned about the science of science communication was more situational or contextual than universal? Or, put more plainly, what if the findings of science communicatio experiments did not apply widely beyond the context they were conducted in?

Let me explain what I mean. We all know that to have a great science experiment means you trial something and then release your findings and methodology, and then you wait for someone to replicate it, demonstrating that your findings were valid.

You may remember back in 1989 the scientific world was full of excitement at the prospect of cold fusion (a type of nuclear reaction taking place at room temperature). This was big news. Almost limitless cheap energy! But the scientists who had made the announcement, Martin Fleischmann and Stanley Pons, went from science media superstars to science media red-faces, when the hundreds of scientists who wanted in on the action found nobody was able to replicate their experiments.[1]

What should we think of all the gold standard science communication theories we've looked at if most of them have difficulty being replicated?

A couple of very significant research projects were released in the last few years, and I expected them to completely rock the underpants off the science communication community. But that didn't happen.

In fact, these studies rarely get much of a mention in the big scheme of things.

The first study was conducted by a collaborative effort of over 270 psychology researchers who got together to try and replicate the findings of 100 key psychological studies. The studies had been published in 2008 in three high-ranking psychology journals (*Psychological Science, Journal of Personality and Social Psychology,* and *Journal of Experimental Psychology: Learning, Memory, and Cognition*). The researchers matched up different experiments to different teams to try and get the best match they could. Then they carefully followed the steps of the original scientists and recruited subjects and administered tests and ran statistical analyses, as closely as they could to the original experiments.[2]

And what did they find? They found they could only replicate about 40% of them – 39 out of the 100 in fact.

The implications of this are pretty profound (to quote *Back to the Future III*), as it has potential impacts across lots of social science research – including science communication research – which is rarely replicated.

But that is the nature of social sciences. Studies are rarely replicated because not only are there no real rewards for replicating other research, but you can actually be subtly punished for it – most often through not achieving publication because your work was deemed 'not new'.

This means that a lot of research that we hold up as a gold standard might really only apply to the situation and context of that time and place, rather than applying universally.

Can you hear the collective uh-oh?

In fact, probing deeper the reproducibility project – as it was known – looked at other key studies and found more problems. One in particular is worth mentioning – a 1999 study entitled 'Stereotype susceptibility: Identity salience and shifts in quantitative performance'. The original study found that:

> ... Asian-American women performed better on a mathematics test when their ethnic identity was activated, but worse when their gender identity was activated, compared with a control group who had neither identity activated ... when a particular social identity was made salient at an implicit level, performance was altered in the direction predicted by the stereotype.[3]

Explained simply, if you prime Asian woman about being Asian (with the associated stereotype of being good at math) they did better at a math test than those primed as being women (who have the stereotype of having low math ability).

That certainly seemed like something worth testing – particularly as it is so often quoted in text books. So two different teams were recruited, one in Georgia and one in California. They replicated the test – and their findings were not the same. The team from Georgia was able to find similar results,[4] but the team from California found that they could not replicate them.[5]

What are we to make of that? Does it reinforce a replication crisis or show that it is only intermittent?

It has certainly spawned a lot of defensive studies in the field of psychology, trying very hard to argue why it is not a crisis if things can't be replicated.[6]

Others have looked deeper at that issue of situation and context and what impact they have on a study. Jay Van Bavel and colleagues, for instance, undertook a study of the reproducibility project's findings, and said many of the studies failed to replicate because it was difficult to recreate, in another time and place, the exact same conditions as those of the original study.[7]

In a comment piece in *The New York Times* Jay Van Bavel wrote:

Imagine a study that examined whether an advertisement for a 'colorblind work environment' was reassuring or threatening to African-Americans. We assumed it would make a difference if the study was conducted in, say, Birmingham, Alabama, in the 1960s or Atlanta in the 2000s.[8]

Did I mention the importance of situation and context? Okay, enough said on that.

The weirdest people in the world

Which brings us to the second study. The key researcher was Joe Henrich, who had been doing work among people in South America in the 1990s. He was an anthropologist from UCLA and rather than do the typical observation thing that most anthropologists do, he decided to try out a behavioural experiment. He took the game known as the prisoner's dilemma to see how the Machiguenga people of Peru that he was working with would react to it.

In the original game there are two players who must remain anonymous to each other. The first player is then given an amount of money, let's say $200. He is told that he has to offer some of that money to the second person. And the second person has the choice of accepting the money or refusing it. But the catch to the game is that if the second player refuses the money, they both go without.

When the game is played in experiments in North America, player one usually offers a 50:50 split, which is accepted. If, however, the divide is less than 50:50 for player two, he or she usually chooses to punish player one by rejecting the money, even though they both miss out.

But among the Machiguenga, the amounts offered by player one were much lower and player two almost never refused the money, no matter how small it was.[9]

What was going on? Didn't the Machiguenga know they were bucking some tenets of basic economic and psychological theory?

Henrich next landed a grant from the MacArthur Foundation to continue his work in other cultures. Further work in Africa and South East Asia found that results continued to differ. He found, for instance, that in societies with a strong emphasis on gift giving to curry favour, player one might offer as much as 60% of the money or more to player two. And player two would still often reject it.

This was something that just wasn't commonly observed in North America.

Let's just pause a moment and consider where most of the social science experiments around the globe are conducted? A 2008 survey of the top six psychology journals found that about 96% of the subjects tested in psychological studies between 2003 to 2007 were Westerners. And about 70% of those come from the United States of America.[9]

If you want to look at it another way, 96% of subjects in these studies come from countries that represent only 12% of the world's population.[9]

Teaming up with two colleagues, Steven J Heine and Ara Norenzayan from the University of British Colombia, the team started applying studies more widely across different cultures, and they found that over and over there was one group of people who were particularly unusual when compared with the broad population of the globe. This group had perceptions, behaviours and motivations that were almost always at one far end of the human bell curve. In recognition of this outlying group they even called their research paper 'The weirdest people in the world'.[10]

And you have probably guessed by now that the weirdest people were North Americans! (With WEIRD also standing for white, educated, industrialised, rich and democratic). And yet they are the main core for global social science experiments.

The researchers stated:

> American participants are exceptional even within the unusual population of Westerners—outliers among outliers.[9]

They concluded that social scientists could not possibly have picked a worse population from which to draw broad generalisations about how we might all behave.

Granted many countries aren't as different to North Americans as they'd like to admit, but for science communicators working in Asia, Africa, the Middle East, South America and the Pacific and so on, the impact of this research is underpants-shaking level concern.

I have sat in many science communication workshops in South East Asia, where the cultures are more collectivist than individualistic, and heard participants quote some of the gold standards of science communication theory – like cultural cognition and backfire and so on, and I've often asked, 'Don't you think you might need to actually test those findings in your own country?'

But testing is expensive and time-consuming while downloading research papers is cheap and relatively quick. What is a person to do?

The Coke and Mentos effect

Individually, the findings of these two research projects are quite startling, but when you mix them together, they are like the Mentos in the Coke bottle effect that all science communicators have tried at one time or another. (If you haven't, go and try it quickly!)

Science communicators really have to start asking ourselves how many of the gold standards of science communication can be applied not just to different times and places, but to different cultures as well. An example is the backfire principle that I cited in Chapter 17. It is the reaction of people, who when they are given facts that show they are wrong in something, dig in even stronger in asserting their own version of the truth. (Try it on any random climate change denier you might know. If they don't want to participate, put some Mentos in their Coke!) The main implication of the backfire effect is that it is not just pointless correcting 'fake news' but it can actually make things worse.[11]

Most of the work on the backfire principle was done by Brendan Nyhan and Jason Reifler, who published a highly-cited paper in 2010 entitled 'When corrections fail: The persistence of political misperceptions'.[12] It has become a fairly well established principle and has had significant impact on how we communicate science facts to people who have their own version of the truth.

And for many science communicators it even became a bit of a mantra: you just can't change people's minds if they disagree on climate change or infant vaccines or gene technology or anything like that. After all, we know that those people have alternative beliefs and alternative facts that support their world views, and they become rusted-on in those beliefs and will defend them in the face of any counter evidence.

Front fire   Back fire

The effects of backfire might not be as much as has been supposed

As Daniel Engber wrote for that great online publication *Slate*, '...it seemed as though America had slipped the moorings of her reason and was swiftly drifting toward a 'post-fact age'.'[13] And he quoted University of Chicago law professor Cass Sunstein, who was one of the first to lament that the internet was making us stupider, back in 2007, saying the internet was clearly to blame for this disaster as online communities served as echo chambers for those with shared beliefs.[14]

But here's another very big question. Did we embrace this study, and many others like it, because it aligned with what we wanted to believe? Did it actually confirm our own beliefs about those 'other people' who are not like us, and who don't believe the things we believe in?

Is it really as simple as – if you tell people one thing, they may well believe the opposite?

Are we, in fact, as guilty in selectively choosing what we wish to believe as those we criticise for believing the things we don't believe in? Because two researchers, Tom Wood and Ethan Porter, undertook a replication of Nyhan and Reifler's study and found that backfire effects may not be so common after all.

Daniel Engber asked the hard question in *Slate*:

> What if this field of research, like so many others in the social sciences, had been tilted toward producing false positive results?[13]

How does that make you feel?

Do you want to deny that is possible? Do you want to say it just doesn't accord with your own experience? Or do you want somebody to prove it before you will accept it?

Well, Wood and Porter recruited over 10 000 subjects and exposed them to corrections to misleading claims made by US political figures from both major parties, on 36 different topics. In fact, they even used the same materials and participants as the Nyhan and Reifler 2010 study. They actually found that on only one issue – the misperception of weapons of mass destruction – was there any backfire. And even then, when the correction was rephrased with simpler wording, no backfire was found.[15]

What they did find was that across all statements, showing people corrections actually moved their beliefs away from the false information. They also found there was an effect between a person's political ideology and the way they viewed statements from different politicians – as did the original study – but they found it wasn't very large. For instance, among US liberals there was an 85% level of correction, amongst moderates 96% and amongst conservatives 83%.

They concluded quite simply:

> The backfire effect is far less prevalent than existing research would indicate.[15]

And they are not the only researchers looking closer at our perhaps mental biases towards studies on people's mental biases.

In 2014, Andy Guess and Alex Coppock from Columbia University looked at a classic 1979 study on the death penalty that found that adding facts to a discussion increased people's disagreement. They looked at the original study question across 683 subjects recruited via the internet, and then went further to test how different kinds of evidence affected the views of another 1170 subjects on the topic of a minimum wage, and then another 2122 people on gun control.

And, you guessed it, in none of these did they find any evidence that people grew more stubborn in their views when presented with disconfirming information.

Instead, they found what Coppock described as 'gorgeous parallel updating'. This meant that people on either side of any issue will actually adjust their beliefs to better fit the facts – not the other way around. He said that backfire was the exception, not the rule, and should be considered a 'truth-y hypothesis' in that it feels right rather than is right.[13]

Other recent studies have found that the echo chamber effect doesn't hold or that news consumption on the internet does not appear to be as fractured as has been thought.[16,17,18]

So what does it all mean?

That's the real very big question, isn't it? Does it mean that everything is so situational and contextual that we can't actually count on the findings of any of the dozens and dozens of studies I've cited in this book unless we replicate them ourselves?

Or does it mean that we need to understand there can always be differences in how research findings might be applied – across all science disciplines – and we should look for the broad underlying principles to apply in our work over the specifics? (Brendan Nyhan and Jason Reifler have been very active in reconsidering their original work to find out just how nuanced and driven by situation and context it can be.[13])

Fortunately, there is an easy answer. We need to apply scientific methods more often in our work, trialling theories before applying them. If many science communication studies are situational and contextual dependent, you need to find ways to apply the findings to your own situations and contexts.

And always be sceptical until proven otherwise.

What to do with what you now know

Clearly having a good grasp of the relevant research around science communication helps – but sometimes you are just going to need to find your own truths.

Even the best and most repeated research might not apply in a particular situation you are involved in. In such instances you need a sharp eye to know when something is not quite working as it should, and have ideas for correcting it to get the best outcomes.

This is particularly true if you are working in countries where little research has been done, as you can't just assume that North American and European studies will play out the same way in your country or region.

That might involve doing a little bit more research of your own. And it might involve talking to other science communicators for ideas.

It's the way the scientific method works. You keep at it and at it and at it, and continually test new ideas, and run trials and find points of impact and then retest them and find ways to scale them up.

No one says this is going to be easy – but if it was you wouldn't need to read a whole book about it, would you?

Key summary points

- Many social studies experiments are not easily replicable.

- The subjects of many social studies experiments are US-undergraduates, who have different beliefs and behaviours to many other people, and make a poor cohort for many studies.

25

Last words: Sermon on the Endnote

'What was that?'
'I think it was "Blessed are the cheesemakers".'

— Monty Python's *The Life of Brian*, 1979

I started this book saying to be an effective science communicator you need three things:

- know your audience
- tell your key message in a good story
- have a clear objective.

However, this book is for a wide variety of audiences and contains many stories. Yet the objective remains simple: I want to distil the breadth of good research that is out there into an easy to understand format.

There is a lot of information in this book, and not all of it is going to be relevant to everyone – but hopefully you will have found the key things that are most relevant to you. For science communication is complex, but the better you understand what research has been undertaken, the better you will be aware of the range of data and tools to help you better communicate science.

I attend a lot of conferences and workshops and, yes even colloquia, that talk about communicating science and the problems and challenges, and lack of impact, and the divide between those who live in the world of science communication practice and those who live in the world of science communication theory. There are few people who are multilingual, as it were, and can move between both of them, but in general people tend to prefer research or practice.

That wouldn't be a bad thing if there were better flows of information between the researchers and the practitioners. But let's be quite honest, a lot of the theory ends up in science journals and is not always easily accessible by those who only speak the language of science practice (and a lot of practitioners don't collect useful data that researchers might be able to use in their work).

I meet many practitioners who say they would like to know more of the research that is printed in academic journals, but in a busy life they don't have the time to pick their way through it all. There are others who are across the research, but feel it has little to offer them in the way of improving their practice. And others still who think a key problem is that science communication researchers are not reaching the many scientists who are out there earnestly trying to communicate science.[1]

Which is a great pity, as there is a wealth of great data in journals that can improve practice in many ways.

But only if it is accessible.

A couple of publications have had a good try at grasping the breadth of academic research and present it in a more accessible form, but I have not been convinced that any of them have really succeeded well in it.

For there is another divide between the currency of published journals and the currency of sharing research and practice widely. And academic publication for many is a currency of personal esteem and of potential promotion and all these other things that are not actually about effectively communicating about science communication.

Now don't get me wrong, academic papers have their place in the world – but they are not the whole world. I have published research papers in several journals, including those that belong to *Nature* and *Cell*, two of the most esteemed journals, with very high impact factors. But to tell you the truth, the sky didn't become bluer, and the kids didn't stop looking at their phones when I talk to them, and my wife didn't put me up on a pedestal – and in fact I didn't even notice people paying much attention to the research before I published it in places like *The Conversation*.

So what is the lesson?

I have read many hundred papers to get the content for this book – but I have also read blogs and articles and online posts and watched YouTube videos, and talked to lots and lots and lots of people. There has been wealth in all of them. I think the real lesson is that if you are doing some really good research, don't wait for someone like me to come along and translate it into plain speak. Also write plain speak articles and blogs and tweets and any things else that science communication practitioners – and just anybody who cares about better communicating science – can read and understand.

And there are so many forms of science communication that are not really covered in this book that also have strong disciples of both practice and theory to learn from on things like:

- data visualisation and graphics
- photography
- videos and animations
- emerging digital technologies
- data mining
- theatre and performance
- event staging
- play-based learning
- education
- museums and science centres.

If these are your thing, you will need to do a bit of digging around to find the data and evidence you need to do them well, but hopefully the structure and content of the chapters in this book will give you a bit of insight into what to find and where you might find it.

Okay – here ends the sermon! Go and do brilliant things.

About the author

Dr Craig Cormick is one of Australia's leading science communicators and has taken part in science communication workshops and activities on all seven continents. He has over 30 years' experience as a science communicator, working predominantly with contentious technologies such as biotechnologies and nanotechnology. He has been widely published on drivers of attitudes to new technologies, including in publications of *Nature* and *Cell*. He has worked for several Australian Government agencies including CSIRO, the Department of Industry, Innovation and Science, and Questacon. He has been President of the Australian Science Communicators and is a member of the Advisory Board on Education and Outreach for the Nobel Prize-winning Organisation for the Prohibition of Chemical Weapons.

Endnotes

Chapter 1

1. Steinbeck J, Rickets J (1941) *Sea of Cortez: A Leisurely Journal of Travel and Research.* Penguin.
2. Roe A (1953) *The Making of a Scientist.* Dodd, Mead, New York.
3. Alda A (2018) *If I Understood You, Would I Have This Look on My Face?* Random House.
4. Nisbet MC, Scheufele D (2009) What's next for science communication? Promising directions and lingering distractions, *American Journal of Botany* **96**(10), 1767–1778.

Chapter 2

1. Jamieson KH, Kahan D, Scheufele D (Eds) (2017) *The Oxford Handbook of the Science of Science Communication.* Oxford University Press, Oxford, UK.
2. Maynard A, Scheufele D (2016) What does research say about how to communicate about science? *The Conversation*, 14 December.
3. National Academies of Sciences, Engineering, and Medicine (2017) *Communicating Science Effectively: A Research Agenda.* The National Academies Press, Washington DC.
4. Dudo A (2012) Toward a model of scientists' public communication activity: The case of biomedical researchers, *Science Communication* **35**(4), 476–501.

Chapter 3

1. Eveland WP, Cooper KE (2013) An integrated model of communication influence on beliefs, *Proceedings of the National Academies of Sciences of the United States of America* **110** (Supplement 3), 14088–14095.
2. National Academies of Sciences, Engineering, and Medicine (2017) *Communicating Science Effectively: A Research Agenda.* The National Academies Press, Washington DC.
3. Kahan D (2013) What is to be done? *Cultural cognition project.* Yale Law School, 19 May. <http://www.culturalcognition.net/blog/2013/5/19/what-is-to-be-done.html>

Chapter 4

1. Doran GT (1981) There's a S.M.A.R.T. way to write management's goals and objectives, *Management Review* **70**(11), 35–36.
2. National Academies of Sciences, Engineering, and Medicine (2017) *Communicating Science Effectively: A Research Agenda.* The National Academies Press, Washington DC.

Chapter 5

1. Cormick C (2014) 'Community attitudes towards science and technology in Australia'. CSIRO, Canberra.
2. Lamberts R (2017) 'The Australian beliefs and attitudes towards science survey'. The Australian National University, Canberra.
3. Ipsos Social Research Institute (2013) 'Community attitudes towards emerging technology issues'. Department of Industry, Innovation, Science, Research and Tertiary Education, Canberra.
4. Bruce G, Critchley C (2012) 'The Swinburne national technology and society monitor 2012.' Faculty of Life & Social Sciences, Swinburne University of Technology, Melbourne, Australia.
5. National Science Board (2016) *Science and Engineering Indicators 2016.* National Science Foundation, Arlington, Virginia.

6. Kennedy B, Funk C (2015) 'Public interest in science and health linked to gender and personality'. Pew Research Center, Washington DC.
7. Funk C, Goo SK (2015) 'A look at what the public knows and does not know about science'. Pew Research Center, Washington DC.
8. Wyatt N, Stolper D (2013) 'Science literacy in Australia'. Australian Academy of Science, Canberra.
9. Hallman WK (2017) What the public think and knows about science – and why it matters. In *The Oxford Handbook of the Science of Science Communication.* (Eds KH Jamieson, D Kahan and DA Scheufele). Oxford Library of Psychology, Oxford University Press, Oxford.
10. National Science Board (2014) *Science and Engineering Indicators 2014.* National Science Foundation, Arlington, Virginia.
11. Pew Research Center (2009) 'Public praises science: scientists fault public media'. Pew Research Center, Washington DC.
12. Pew Research Center (2015) 'Public and scientists' views on science and society'. Pew Research Center, Washington DC.

Chapter 6

1. Dubner S, Levitt S (2005) *Freakonomics,* Harper Collins.
2. Ford H, Crowther S (1922) *My Life and Work,* Doubleday.
3. Aldoory L, Grunig JE (2012) The rise and fall of hot-issue publics: relationships that develop from media coverage of events and crises, *International Journal of Strategic Communication* 6(1), 93–108.
4. Leiserowitz A, Maibach E, Roser-Renouf C, Feinberg G, Rosenthal S (2015) 'Global warming's six Americas.' March 2015. Yale Program on Climate Change Communication, Yale University and George Mason University, New Haven, Connecticut.
5. Yale Program on Climate Change Communication & George Mason University's Center for Climate Change Communication. <http://climatecommunication.yale.edu/about/projects/global-warmings-six-americas/>.
6. Lim-Camacho L, Ariyawardana A, Lewis G, Crimp S (2014) 'Climate adaptation: What it means for Australian consumers. Consumer survey – 2014 results'. CSIRO, Canberra.
7. Nielsen (2014) 'Report on public attitudes towards science and technology.' Nielsen, New Zealand.
8. Funk C, Goo SK (2015) 'A look at what the public knows and does not know about science'. Pew Research Center, Washington DC.
9. Ipsos-Mori (2011) 'Public attitudes to science'. Department for Business, Innovation & Skills, UK.
10. Cormick C (2014) 'Community attitudes towards science and technology in Australia'. CSIRO, Canberra.
11. Cormick C (2014) Social research into public attitudes towards new technologies, *Journal für Verbraucherschutz und Lebensmittelsicherheit* 9 (Supplement 1), 39–45.
12. Cormick C, Romanach LM (2014) Segmentation studies provide insights to better understanding attitudes towards science and technology, *Trends in Biotechnology* 32(3), 114–116.
13. McCrindle Research (2018) <www.mccrindle.com.au>.

Chapter 7

1. De Broglie L (1962) *New Perspectives in Physics.* Basic Books, New York.
2. National Academies of Sciences, Engineering, and Medicine (2017) *Communicating Science Effectively: A Research Agenda.* The National Academies Press, Washington DC.
3. Fischhoff B, Brewer NT, Downs JS (2011) *Communicating Risks and Benefits: An Evidence-Based User's Guide.* FDA, Department of Health and Human Services, Silver Spring.
4. Committee on Public Understanding of Engineering Messages (2008) *Changing the Conversation: Messages for Improving Public Understanding of Engineering.* The National Academies Press, Washington DC.

5. Huertas A (2016) Developing Effective Messages in Science Communication, *Science Communication Media*, 30 June. <https://medium.com/science-communication-media/developing-effective-messages-in-science-communication-9b658df3e672>

6. COMPASS Science Communication, Inc. (2017) *The Message Box Workbook.* <https://www.compassscicomm.org/>

7. Andrei M (2016) Alan Alda's important message for science communication. *ZMEScience*, 15 March. < https://www.zmescience.com/science/alan-aldas-important-message-for-science-communication/>

8. Shaha A (2008) Credit where credit is due. *Lablit*, 30 March. < http://www.lablit.com/article/365>

9. Scharf CA (2013) In defence of metaphors in science writing. *Scientific American,* 9 July. < https://blogs.scientificamerican.com/life-unbounded/in-defense-of-metaphors-in-science-writing/>

10. Pauwels E (2013) Communication: Mind the metaphor, *Nature* **500**(7464), 523–524.

11. Khakhar A (2017) Analogies and metaphors in science communication: the good and the bad. *Engage*, 19 January. <https://courses.washington.edu/engageuw/analogies-and-metaphors-in-science-communication-the-good-and-the-bad/>

12. Kueffer C, Larson BMH (2014) Responsible use of language in scientific writing and science communication. *BioScience* **64**(8), 719–724.

13. Goodenough U (2010) Gravity is love, and other astounding metaphors, *Cosmos and Culture*, National Public Radio, 21 October. < https://www.npr.org/sections/13.7/2010/10/21/130724690/gravity-is-love>

14. Costandi M (2013) Mo Costandi on science writing: a good story conveys wonderment. *The Guardian*, 22 April. < https://www.theguardian.com/science/2013/apr/22/mo-costandi-science-writing>

15. Thibodeau PH, Boroditsky L (2011) Metaphors we think with: the role of metaphors in reasoning. *PLoS ONE* **6**(2), e16782.

16. Flushberg S, Matlock T, Thibodeau PH (2016) Metaphors for the war (or race) against climate change. *Journal of Environmental Communication* **11**(6), 769–783.

17. Aurbach E, Prater KE, Patterson B, Zikmund-Fisher BJ (2018) Half-life your message: a quick, flexible tool for message discovery, *Science Communication* **40**(5), 669–677.

Chapter 8

1. Dennison B (2016) Evolution's hero vs. a historical footnote: A new narrative Index sheds light on Darwin vs. Wallace. *Integration and Application Network*, 24 February. <http://ian.umces.edu/blog/2016/02/24/evolutions-hero-vs-a-historical-footnote-a-new-narrative-index-sheds-light-on-darwin-vs-wallace/>

2. Olson R (2015) *Houston, We Have a Narrative.* University of Chicago Press, Chicago.

3. Huertas A (2015) Book review: Houston, We Have a Narrative by Randy Olson. *Union of Concerned Scientists*, 14 October. < https://blog.ucsusa.org/aaron-huertas/book-review-houston-we-have-a-narrative-by-randy-olson-923>

4. Reagan AJ, Mitchel L, Kiley D, Danforth CM, Dods PS (2016) The emotional arcs of stories are dominated by six basic shapes. *EPJ Data Science* **5**:31.

5. Fischhoff B, Brewer NT, Downs JS (2011) *Communicating Risks and Benefits: An Evidence-Based User's Guide.* FDA, Department of Health and Human Services, Silver Spring.

6. Montgomery S L (2003) *The Chicago Guide to Communicating Science.* University of Chicago Press, Chicago.

7. Sheehan M, Christiano A, Neimand A (2018) Science of story building: true stories also have to feel true—verisimilitude. *The Science of Story Building*, 7 May. < https://medium.com/science-of-story-building/identifying-verisimilitude-258eaab15ee>

8. Niemand A (2018) How to tell stories about complex issues. *Stanford Social Innovation Review*, 7 May. < https://ssir.org/articles/entry/how_to_tell_stories_about_complex_issues>

9. Dahlstrom M (2014) Using narratives and storytelling to communicate science with non-expert audiences. *Proceedings of the National Academies of Science*, September 16.

10. Downs JS (2014) Prescriptive scientific narratives for communicating usable science. *Proceedings of the National Academies of Science*, **111** (Supplement 4), 13614–13620.

11. National Academies of Sciences, Engineering, and Medicine (2017) *Communicating Science Effectively: A Research Agenda*. The National Academies Press, Washington DC.

12. Krauss LM (2016) Finding beauty in the darkness. *The New York Times*, Section Sunday Review, 11 February.

13. Forster EM (1927) *Aspects of the Novel*. Harcourt, Brace & Company, New York.

14. Winterbottom A, Bekker HL, Conner M, Mooney A (2008) Does narrative information bias individual's decision making? A systematic review. *Social Science & Medicine* **67**(12), 2079–2088.

15. Munshi D, Kurian P (2018) Framing futures through fiction and folklore: weaving past and prospective narratives of public understanding of climate science, Public Communication of Science and Technology, University of Dunedin. (conference presentation)

16. Kaplan M (2013) The narratives of science communication. In *Science of Science Communication II*, 23–25 September, National Academy of Sciences, Washington DC.

17. Graesser AC, Hauft-Smith K, Cohen AD, Pyles LD (2015) Advanced outlines, familiarity, and text genre on retention of prose. *The Journal of Experimental Education* **48**(4), 281–290.

18. Green MC, Brock TC (2000) The role of transportation in the persuasiveness of public narratives. *Journal of Personality and Social Psychology.* **79**(5), 701–721.

Chapter 9

1. Roy Morgan (2017) *Images of Professions Survey 2017*. < http://www.roymorgan.com/findings/7244-roy-morgan-image-of-professions-may-2017-201706051543>

2. Brenan M (2017) Nurses keep healthy lead as most honest, ethical profession. Gallup, 26 December. <https://news.gallup.com/poll/224639/nurses-keep-healthy-lead-honest-ethical-profession.aspx

3. Korn Group (2017) The Trust Edition, Korn Group. <http://thekorngroup.com.au/>

4. Smith C (2013) The 100 most trusted people in America. *Readers Digest*, <https://www.rd.com/culture/readers-digest-trust-poll-the-100-most-trusted-people-in-america/1/>

5. Hoffman S, Tan C (2013) Following celebrities' medical advice: meta-narrative analysis, *British Medical Journal* **347**, f7151.

6. Bellhuz J, Hoffman S (2013) Katie Couric and the celebrity medicine syndrome, *Los Angeles Times*, 18 December.

7. Waxman O (2015) This is the world's hottest accent. *Time*, 10 February. <http://time.com/3702961/worlds-hottest-accent/>

8. McClatchy-Marist (2017) <https://www.documentcloud.org/documents/3532285-McClatchy-Marist-Poll-National-Nature-of-the.html>

9. Fahy D (2015) *The New Celebrity Scientists. Out of the Lab and Into the Limelight*. Rowman & Littlefield Publishers, London.

10. Ipsos-Mori (2017) *Trust in Professions*. Ipsos-Mori. <https://www.ipsos.com/sites/default/files/ct/news/documents/2017-11/trust-in-professions-veracity-index-2017-slides.pdf>

11. Searle SD (2014) How do Australians engage with science? Preliminary results from a national survey. Australian National Centre for the Public Awareness of Science (CPAS), The Australian National University, Canberra.

12. National Science Board (2016) *Science and Engineering Indicators 2016*. National Science Foundation, Arlington, Virginia.

13. Pew Research Center (2016) Public opinion about genetically modified foods and trust in scientists connected with these foods. <http://www.pewinternet.org/2016/12/01/public-opinion-about-genetically-modified-foods-and-trust-in-scientists-connected-with-these-foods/>

14. Pew Research Center (2015) 'Americans, politics and science issues.' Pew Research Center, Washington DC.

15. Dastagir A (2017) People trust science. So why don't they believe it? *USA Today*, 2 June.

16. Brewer PR, Ley BL (2013) Whose science do you believe? Explaining trust in sources of scientific information about the environment. *Science Communication* **35**(1), 147–173.

17. Edelman (2018) *2018 Edelman Trust Barometer*. Edelman. < https://www.edelman.com/research/2018-edelman-trust-barometer>

18. Resnick HE, Sawyer K, Huddleston N (2015) *Trust and Confidence at the Interfaces of the Life Sciences and Society: Does the Public Trust Science? A Workshop Summary*. The National Academies Press, Washington DC.

19. Rabinovich A, Morton TA (2012) Unquestioned answers or unanswered questions: Beliefs about science guide responses to uncertainty in climate change risk communication. *Risk Analysis* **32**(6), 992–1002.

20. Slovic P (1999) Trust, emotion, and sex. *Risk Analysis* **19**(4), 689–701.

21. Hon LC, Grunig JE (1999) Guidelines for Measuring Relationships in Public Relations, Institute for Public Relations, Florida.

22. Czerski H (2017) A crisis of trust is looming between scientists and society – it's time to talk. *The Guardian*, 27 January. < https://www.theguardian.com/science/blog/2017/jan/27/a-crisis-of-trust-is-looming-between-scientists-and-society-its-time-to-talk>

23. Kipnis D (1996) Trust and technology. In *Trust in Organizations: Frontiers of Theory and Research*. (Eds R Kramer and T Tyler) Sage Publications, California.

24. Willinghame D (2011) Trust me I'm a scientist. *Scientific American*, 1 May. < https://www.scientificamerican.com/article/trust-me-im-a-scientist/>

25. Greenfield S (2010) Trust me, I'm a scientist. ABC Science, 16 June. < http://www.abc.net.au/science/articles/2010/06/16/2928357.htm>

Chapter 10

1. Eliott S (2018) email communication, 10 October.

2. O'Connell A, Greene CM (2016) Not strange but not true: self-reported interest in a topic increases false memory. *Memory*, **25**(8), 969–977.

3. Williams R (2018) *Turmoil: Letters from the Brink*. New South Publishing, Sydney.

4. Porter E (2015) Science communication: Science in the media. *Nature Jobs Blog*, 8 July. < http://blogs.nature.com/naturejobs/2015/07/08/science-communication-science-in-the-media/>

5. Nisbet MC, Mooney C (2007) Framing science. *Science* **316**, 56.

6. Sumner P, Vivian-Griffiths S, Boivin J, Williams A, Venetis CA, Davies A, Ogden J, Whelan L, Hughes B, Daltonm B, Boy F, Chambers CD (2014) The association between exaggeration in health related science news and academic press releases: retrospective observational study. *British Medical Journal* **349**, g7015.

7. Woloshin S, Schwartz LM, Casella SL (2009) Press releases by academic medical centers: Not so academic? *Annals of Internal Medicine* **150**(9), 613–618.

8. Nisbet MC (2009) Framing science: A new paradigm in public engagement. In *New Agendas in Science Communication*. (Eds LeeAnne Kahlor and Patricia Stout) Taylor & Francis Publishers.

9. Ellerton P (2014) The problem of false balance when reporting on science. *The Conversation*, 17 July.

10. Mitchell A, Funk C, Gottfried J (2017) 'Science news and information today'. Pew Research Centre, 20 September. <http://www.journalism.org/2017/09/20/science-news-and-information-today/>

11. Sandman P (2001) Explaining environmental risk. *The Peter M. Sandman Risk Communication Website*. <http://www.psandman.com/articles/explain3.htm>

12. Castell S, Charlton A, Clemence M, Pettigrew N, Pope S, Quigley A, Shah JN, Silman T (2014) 'Public attitudes to science'. Ipsos-Mori Social Research Unit for Department for Business, Innovation & Skills, UK.

13. Pew Research Center (2017) 'Science news and information today'. Pew Research Centre, Washington DC.

14. Cormick C (2014) 'Community attitudes towards science and technology in Australia'. CSIRO, Canberrra.
15. Yeo SK, Brossard D (2017) The (changing) nature of scientist-media interactions: a cross-national analysis. In *The Oxford Handbook of the Science of Science Communication*. (Eds KH Jamieson, D Kahan and D Scheufele) Oxford University Press, Oxford.
16. Cormick C, Mercer R (2017) 'Community attitudes to gene technology'. Report prepared for the Office of the Gene Technology Regulator. Instinct and Reason, Sydney Australia.
17. Gluckman P (2017) Can science and science advice be effective bastions against the post-truth dynamic? STS series, University College London, 18 October.
18. Peters HP, Brossard D, de Cheveigne S, Dunwoody S, Kallfass M, Miller S, Tsuchida S (2008) Science media interface: It's time to reconsider, *Science Communication*, **30**(2), 266–276.
19. Science Media Centre (2002) 'MMR: Learning lessons. A report on the meeting hosted by the Science Media Centre, 2 May 2002.' Science Media Centre, London.
20. Li N, Lull RB (2017) A recap: The role, power and peril of media for the communication of science. In *The Oxford Handbook of the Science of Science Communication*. (Eds KH Jamieson, D Kahan and D Scheufele) Oxford University Press, Oxford.
21. Jackson JK, Mahar I, Gaultois M, Altosaar J (2016) Accurate science or accessible science in the media – why not both? *The Conversation*, 2 June.
22. Guidelines for scientists on communicating with the media, Social Issues Research Centre and Amsterdam School of Communications Research. <http://www.sirc.org/messenger/>
23. Porter E (2015) Science communication: Science in the media. *Nature Jobs Blog*, 8 July. < http://blogs.nature.com/naturejobs/2015/07/08/science-communication-science-in-the-media/>
24. Press release guidelines for scientists. Hubble Space Telescope, The Hubble European Space Agency Information Centre (undated). < https://www.spacetelescope.org/about_us/scientist_guidelines/>
25. Bearup G (2017) Of mice and men. *The Weekend Australian Magazine*, 25–26 November.
26. Young E (2013) A guide for scientists on giving comments to journalists, *National Geographic*, 22 May.
27. Eliott S (2009) Controversy: Silence is a scientist's worst enemy. *Issues* **87**, June.

Chapter 11
1. Harris S (2018) Sam Harris discusses Donald Trump's rewriting of reality and how social media drives us insane. *ABC Online*, 1 June. < https://www.abc.net.au/news/programs/the-world/2018-05-31/sam-harris-discusses-donald-trumps-rewriting-of/9822642>
2. Booth M (2016) Science communication vital in post-truth world. *The Australian*, 30 November.
3. Yeo SK, Brossard D (2017) The (changing) nature of scientist-media interactions: a cross-national analysis. In *The Oxford Handbook of The Science of Science Communication*. (Eds KH Jamieson, D Kahan and D Scheufele) Oxford University Press, Oxford.
4. Smith A, Anderson M (2018) 'Social media use in 2018.' Pew Research Center, Washington DC.
5. Cowling D (2018) Social Media Statistics Australia – April 2018. *SocialMediaNews.com.au*, 1 May. < https://www.socialmedianews.com.au/social-media-statistics-australia-april-2018/>
6. Kemp S (2018) Digital in 2018: World's internet users pass the 4 billion mark. We are social. 30 January. <https://wearesocial.com/blog/2018/01/global-digital-report-2018>
7. Chaffey D (2019) Global social media research summary 2019. Smart Insights, 12 February. <https://www.smartinsights.com/social-media-marketing/social-media-strategy/new-global-social-media-research/>
8. Dwivedi S (2017) Communicating science through social media. *Technical Today*, 18 November. < http://technicaltoday.in/communicating-science-through-social-media/>
9. Kenkel B (2017) Social media as a scientist: a very quick guide. *Nature Jobs Blog*, 23 August. < http://blogs.nature.com/naturejobs/2017/08/23/social-media-as-a-scientist-a-very-quick-guide/>
10. Gluckman P (2017) Can science and science advice be effective bastions against the post-truth dynamic? STS series, University College London, 18 October.

11. Sloman S, Fernback P (2017) *The Knowledge Illusion: Why we Never Think Alone*, Penguin-Random House.

12. Scheufele DA (2014) Science communication as political communication. *Proceedings of the National Academy of Sciences,* **111** (Supplement 4), 13585–13592.

13. Bik HM, Goldstein MC (2013) An introduction to social media for scientists. *PLoS Biology* **11**(4), e1001535.

14. Terras M (2012) Is blogging and tweeting about research papers worth it? The verdict. *Melissa Terras' Blog.* 3 April. < http://melissaterras.blogspot.com/2012/04/is-blogging-and-tweeting-about-research.html>

15. Eysenbach G (2011) Can tweets predict citations? Metrics of social impact based on twitter and correlation with traditional metrics of scientific impact. *Journal of Media Internet Research,* **13**(4), e123.

16. Australian Science Media Centre, Tips for scientists using social media, *Science Media Savvy.* <http://sciencemediasavvy.org/using-social-media/>

17. Brown JP (2017) 3 secrets to social media for science communication. *From the Lab Bench,* 15 January. <http://www.fromthelabbench.com/from-the-lab-bench-science-blog/2017/1/15/3-secrets-to-social-media-for-science-communication>

18. National Academy of Sciences (2018) *The Science of Science Communication III: Inspiring Novel Collaborations and Building Capacity: Proceedings of a Colloquium.* The National Academies Press, Washington DC.

19. Rubel S (2018) Truth is not self-evident, we must make it so. Edelman, 21 January. < https://www.edelman.com/post/truth-is-not-self-evident>

20. Mollett A, Brumley, Gilso C, Williams S (2017) *Communicating Your Research with Social Media: A Practical Guide to Using Blogs, Podcasts, Data Visualisations and Video,* Blackwells.

21. Mollett A, Brumley C, Gilson C, Williams S (2018) Social media and the research lifestyle. Sage Publishing. <https://study.sagepub.com/mollett2/student-resources/chapter-1/social-media-and-the-research-lifestyle>

22. Lunt I (2015) Live tweeting at academic conferences: time to move on? *Ian Lunt Ecology,* 17 December. <https://ianluntecology.com/2015/12/17/live-tweeting-at-academic-conferences/>

23. Mangoo F (2013) You won't finish this article: Why people online don't read to the end. *Slate,* 6 June. < https://slate.com/technology/2013/06/how-people-read-online-why-you-wont-finish-this-article.html>

24. Lunt I (2014) Never blog in your PJs, and other tips for science and ecology bloggers. *Ian Lunt Ecology,* 27 October. <https://ianluntecology.com/2014/10/27/never-blog-in-pjs/>

25. AFM Radio (2018) Zambia, 4 April.

26. Schulson M (2018) Are Google and Facebook responsible for the medical quackery they host? *Undark,* 6 June.

27. Chigwedere P, Seage GR, Gruskin S, Lee TH, Essex M (2008) Estimating the lost benefits of antiretroviral drug use in South Africa. *Journal of Acquired Immune Deficiency Syndrome* **49**(4), 410–415.

28. MacInnes P (2018) What's up PewdiePie? The troubling content of YouTube's biggest star, *The Guardian,* 5 April. < https://www.theguardian.com/tv-and-radio/2018/apr/05/whats-up-pewdiepie-the-troubling-content-of-youtubes-biggest-star>

29. Ballance A (2018) Science film-maker a winner. Our Changing World, Radio New Zealand, 15 February.

30. Amarasekara I, Grant W (2018) Exploring the YouTube science communication gender gap: a sentiment analysis, *Public Understanding of Science,* **28**(1), 68–84.

31. Tsou A, Thelwall M, Mongeon P, Sugimoto CR (2014) A community of curious souls: an analysis of commenting behaviour on TED talks videos. *PLoS ONE,* **9**(4), e93609.

32. Jeffries A (2018) Women making science videos on YouTube face hostile comments. *The New York Times,* 13 July.

33. Welbourne D, Grant W (2015) What makes a popular science video on YouTube. *The Conversation*, 25 February.

Chapter 12

1. Druckman JN, Bolsen T (2011) Framing, motivated reasoning, and opinions about emergent technologies. *Journal of Communication* **61**(4), 659–688.
2. Lakoff G (2004) *Don't Think of an Elephant! Know your Values and Frame the Debate*, Chelsea Green Publishing, Vermont.
3. Cormick C (2012) Ten big questions on public engagement on science and technology: observations from a rocky boat in the upstream and downstream of engagement. *International Journal of Deliberative Mechanisms in Science* **1**(1), 35–50.
4. Nisbet MC, Scheufele DA (2009) What's next for science communication? Promising directions and lingering distractions. *American Journal of Botany* **96**(10), 1767–1778.
5. Van der Linden S, Maibach E, Leiserowitz A (2015) Improving public engagement with climate change: five 'Best Practice' insights from psychological science. *Perspectives on Psychological Science* **10**(6), 758–763.
6. Lakoff G (2010) Why it matters how we frame the environment. *Environmental Communication* **4**(1), 70–81.
7. Nisbet MC (2009) Framing science: a new paradigm in public engagement. In *New Agendas in Science Communication*. (Eds LeeAnne Kahlor and Patricia Stout) Taylor & Francis Publishers.
8. McCright AM, Dunlap RE (2003) Defeating Kyoto: The conservative movement's impact on US climate change policy. *Social Problems* **50**(3), 348–373.
9. Boykoff M, Boykoff J (2004) Balance as bias: global warming and the US prestige press. *Global Environmental Change* **14**, 125–136
10. Tversky A, Kahneman D (1986) Rational choice and the framing of decisions. *The Journal of Business*, **59**(4), Part 2: The Behavioral Foundations of Economic Theory, S251–S278.
11. Nisbet MC (2009) Communicating climate change: why frames matter for public engagement. *Environment: Science and Policy for Sustainable Development* **51**(2), 12–23.
12. National Academies of Sciences, Engineering, and Medicine (2017) *Communicating Science Effectively: A Research Agenda*. The National Academies Press, Washington DC.
13. Bruine de Bruin W, Wong-Parodi G (2014) The role of initial affective impressions in responses to educational communications: The case of carbon capture and sequestration (CCS). *Journal of Experimental Psychology: Applied* **20**(2), 126–135.
14. Nisbet MC, Mooney C (2007) Framing science. *Science and Society* **316**(5821), 56.
15. Nisbet MC, Kotcher J (2009) A two step flow of influence? Opinion-leader campaigns on climate change. *Science Communication* **30**(3), 328–354.
16. Druckman J, Lupia A (2017) Using frames to make scientific communication more effective. In *The Oxford Handbook of the Science of Science Communication* (Eds KH Jamieson KH, D Kahan and D Scheufele) Oxford University Press, Oxford.

Chapter 13

1. Brewer G (2001) Snakes top list of Americans' fears. Gallup, 19 March. <https://news.gallup.com/poll/1891/snakes-top-list-americans-fears.aspx>
2. Chapman University Survey of American Fears (2018) America's top fears 2018. Chapman University, 11 October. <https://blogs.chapman.edu/wilkinson/2018/10/16/americas-top-fears-2018/>
3. Kangas Dwyer K, Davidson MM (2012) Is public speaking really more feared than death? *Communication Research Reports* **29**(2), 99–107.
4. Croston G (2012) The thing we fear more than death: why predators are responsible for our fear of public speaking. *Psychology Today*, 29 November. <https://www.psychologytoday.com/au/blog/the-real-story-risk/201211/the-thing-we-fear-more-death>

5. Treise D, Weigold MF (2002) Advancing science communication: a survey of science communicators. *Science Communication* **23**(3), 310–322.

6. Flaxington B (2015) Overcoming fear of public speaking. *Psychology Today*, 16 March. <https://www.psychologytoday.com/au/blog/understand-other-people/201503/overcoming-fear-public-speaking>

7. Jackson B, Compton J, Thornton AL, Dimmock JA (2017) Re-thinking anxiety: using inoculation messages to reduce and reinterpret public speaking fears. *PLoS ONE* **12**(1), e0169972

8. Sawchuk CN (2017) Fear of public speaking: How can I overcome it? Mayo Clinic, 17 May. < https://www.mayoclinic.org/diseases-conditions/specific-phobias/expert-answers/fear-of-public-speaking/faq-20058416>

9. Kim S (2016) 7 public speaking tips from researchers who studied 100,000 presentations. ABC News, 19 April. < https://abcnews.go.com/Business/public-speaking-tips-researchers-studied-100000-presentations/story?id=38514076>

10. Johnson R, Johnson N (2016) How to overcome your fear of public speaking. British Council, 10 October. <https://www.britishcouncil.org/voices-magazine/how-overcome-fear-public-speaking>

11. Antonakis J, Fenley M, Liechti S (2012) Learning charisma. *Harvard Business Review*, June.

12. Olson R (2018) *Don't Be Such a Scientist: Talking Substance in an Age of Style*. Island Press, Centre for Resource Economics, Washington DC.

13. Duarte N (2010) *Resonate: Present Visual Stories that Transform Audiences*. John Wiley & Sons Inc.

14. Van Edwards V (2017) *Captivate: The Science of Succeeding with People*. Random House.

15. Kelly S (2012) Gestures fulfil a big role in language. Science Daily, 8 May. <https://www.sciencedaily.com/releases/2012/05/120508152000.htm>

16. 9 simple and effective public speaking tips for scientists. Scientifica *NeuroWire*. <https://www.scientifica.uk.com/neurowire/9-simple-and-effective-public-speaking-tips-for-scientists>

17. Orzel C (2013) The quirks of scientific public speaking. chadorzel.com, 11 June. <http://chadorzel.com/principles/2013/06/11/the-quirks-of-scientific-public-speaking/>

18. Berkun S (2013) Why do people make bad slides? www.scottberkun.com, 6 June. <https://scottberkun.com/2009/why-do-people-make-bad-slides/>

19. Atherton C (2009) Visual attention: a psychologist's perspective. www.slideshare.net, 29 September. <https://www.slideshare.net/CJAtherton/chris-atherton-at-tcuk09>

20. Rogers S (2013) John Snow's data journalism: the cholera map that changed the world. *The Guardian*, 15 March. <https://www.theguardian.com/news/datablog/2013/mar/15/john-snow-cholera-map>

21. Brasseur L (2005) Florence Nightingale's visual rhetoric in the rose diagrams, *Technical Communication Quarterly* **14**(2), 161–182.

22. Couron A (2017) The 9 worst data visualisations ever created. Living QlikView, 2 May. <http://livingqlikview.com/the-9-worst-data-visualizations-ever-created/>

23. Sharma N (2015) 7 most common data visualisation mistakes. thenextweb.com, 16 May. < https://thenextweb.com/dd/2015/05/15/7-most-common-data-visualization-mistakes/>

Chapter 14

1. Centers for Disease Control and Prevention (2011) *Principles of Community Engagement*. 2nd edn. Department of Health and Human Services, USA.

2. Chappell B (2008) *Community Engagement Handbook: A Model Framework for leading practice In Local Government in South Australia*. Government of South Australia and Local Government Association of South Australia, Adelaide.

3. Lane T, Hicks J (2014) *Best Practice Community Engagement in Wind Development*. Commissioned by the ACT Government Environment and Planning Directorate, Canberra.

4. The IAP2 Spectrum of Public Participation, IAP2. <https://www.iap2.org.au/About-Us/About-IAP2-Australasia-/Spectrum>

5. Select Committee on Science and Technology (2001) *Science and Society. Third Report of the Session 1999–2000*. House of Lords, UK Parliament.

6. Von Schomberg R, Davies S (2010) *Understanding Public Debate on Nanotechnologies: Options for Framing Public Policy*. European Commission Services, Belgium.
7. Gilbert J (2007) 'Community education, awareness and engagement programs for bushfire: an initial assessment of practices across Australia'. Technical report C0701, Bushfire CRC, Melbourne.
8. Paton D (2009) Witness statement of Douglas Paton to the Victorian Bushfires Royal Commission. Issued 16 February 2009. <http://www.royalcommission.vic.gov.au/getdoc/5bcf161b-93c2-4a63-bd1d-f3f99ba42648/WIT.031.001.0001.pdf>
9. Cormick C (2012) Ten big questions on public engagement on science and technology: observations from a rocky boat in the upstream and downstream of engagement. *International Journal of Deliberative Mechanisms in Science* **1**(1), 35–50.
10. Department of Fire and Emergency Services (2014) Community engagement framework. Western Australian Government, Perth.
11. Russell AW (2013) Improving legitimacy in nanotechnology policy development through stakeholder and community engagement: forging new pathways. *Review of Policy Research* **30**(5), 566–587.
12. Rowe G, Frewer L (2000) Public participation methods: A framework for evaluation. *Science and Technology of Human Values* **25**(1), 3–29.
13. Abels G (2006) Forms and functions of participatory technology assessment – Or: Why should we be more sceptical about public participation? In *Proceedings of Participatory Approaches in Science and Technology Conference*, 4–7 June, Edinburgh, Scotland.
14. <https://scistarter.com/citizenscience.html>
15. Davis J, Ritchie E, Martin J, Maclagan S (2016) The rise of citizen science is great news for our native wildlife. *The Conversation*, 17 August.
16. Norris R (2017) Exoplanet discovery by an amateur astronomer shows the power of citizen science. *The Conversation*, 7 April.
17. Aceves-Bueno E, Adeleye A, Feraud M, Huang Y, Tao M, Yang Y, Anderson S (2017) The accuracy of citizen science data: a quantitative review. *Bulletin of the Ecological Society of America* **98**(4), 278–290.
18. Golumbic YN (2015) What makes citizen science projects successful, and what can we learn from them for future projects? A literature review of citizen science projects. Technion Citizen Science Project, Israel.
19. Clyde ME (2015) 'Tips for working with Cctizen science volunteers'. NH Cooperative Extension Specialist, Community Volunteer Development, University of New Hampshire.
20. Australian Institute for Disaster Resilience (2010) *Guidelines for the Development of Communication Education, Awareness and Engagement Programs*. Manual 45. Australian Institute for Disaster Resilience, Melbourne.

Chapter 15

1. Dahlstrom MF, Ho SS (2012) Ethical considerations of using narrative to communicate science. *Science Communication* **34**(5), 592–617.
2. Scheufele DA (2014) Science communication as political communication. *Proceedings of the National Academy of Sciences*, **111** (Supplement 4), 13585–13592.
3. Axelrod A (2009) *The Real History of the Cold War: A New Look at the Past*. Sterling, New York.
4. Diamond J (2005) *Collapse: How Societies Choose to Fail or Succeed*. Viking Press.
5. National Academies of Sciences, Engineering, and Medicine (2017) *Communicating Science Effectively: A Research Agenda*. The National Academies Press, Washington DC.
6. Maynard A, Scheufele D (2016) What does research say about how to effectively communicate about science? *The Conversation*, 14 December.
7. Hillman N (2016) The 10 commandments for influencing policymakers, *Times Higher Education*, 26 May.

8. Cairney P, Kwiatkowski R (2017) How to communicate effectively with policymakers: combine insights from psychology and policy studies, *Palgrave Communications* **3**, article number 37.

9. Chubb I (2018) Panel comment at Canberra Writers Festival, 26 August.

10. D'Agostino J (2015) Policy briefs. Centre for the Implementation of Public Policies Promoting Equity and Growth, Global Development Network.

11. Arkeny R (2018) Science meets Parliament doesn't let the rest of us off the hook. *The Conversation*, 12 February.

12. Science and Technology Australia (2018) Science meets Parliament: creating vision beyond party and policy. scienceandtechnologyaustralia.org.au, 2 February. <https://scienceandtechnologyaustralia.org.au/science-meets-parliament-creating-vision-beyond-party-and-policy/>

13. Weyrauch V, Echt L, Arrieta D (2013) How to communicate research for policy influence. Toolkit No.1: First approach to research communication. Buenos Aires. CIPPEC.

14. Anderson S (2017) Victorian brumbies: invasive pest, or majestic part of our heritage? ABC News. 29 January.

15. Office of Environment and Heritage (2016) 'Draft wild horse management plan, Kosciuszko National Park'. NSW Government Office of Environment and Heritage, Sydney.

16. Hull C (2017) New info networks v old political hierarchy. *The Canberra Times*, 30 December.

17. Nisbet MC, Scheufele DA (2009) What's next for science communication? Promising directions and lingering distractions. *American Journal of Botany* **96**(10), 1767–1778.

Chapter 16

1. Jensen E (2014) The problems with science communication evaluation. *Journal of Science Communication* **13**(01), C04.

2. Robinson TN, Patrick K, Eng TR, Gustafson D (1998) An evidence-based approach to interactive health communication: A challenge to medicine in the information age. *Journal of the American Medical Association* **280**(14), 1264–1269.

3. M Davies, Heath C (2013) Evaluating evaluation: increasing the impact of summative evaluation in museums and galleries. King's College London.

4. Edwards C (2004) Evaluating European public awareness of science initiatives. *Science Communication* **25**(3), 260–271.

5. Jensen EA (2015) Evaluating impact and quality of experience in the 21st century: using technology to narrow the gap between science communication research and practice. *Journal of Science Communication* **14**(03), C05

6. Elliott J, Longnecker N (2013) *Inspiring Australia Evaluation Resources*. Inspiring Australia, Canberra.

7. Joubert M (2007) Evaluating science communication projects. SciDevNet, 1 August. <https://www.scidev.net/global/communication/practical-guide/evaluating-science-communication-projects-1.html>

8. Jensen EA (2015) Highlighting the value of impact evaluation: enhancing informal science learning and public engagement theory and practice. *Journal of Science Communication* **14**(03), Y05.

9. Gardiner HR (2017) Learnings From #uwescu17 #2: Evaluating science communication, Heidi R Gardner blog, 21 December. <https://heidirgardner.com/2017/12/21/learnings-from-uwescu17-1-evaluating-science-communication/>

10. Negrete A, Lartigue C (2010) The science of telling stories: evaluating science communication via narratives (RIRC method). *Journal Media and Communication Studies* **2**(4), 98–110.

11. Cole S (2018) Evaluation: expensive, undervalued, ethereal? *Econnect newsletter*, February. <http://www.econnect.com.au/02/2018/february-2018-questions-science-communication-practitioners-want-answered-by-people-researching-science-communication/>

12. Falk J, Needham M, Dierking L, Prendergast L (2014) 'International Science Centre impact study'. John H. Falk Research, Oregon, USA.
13. Modified from Tyler Vigen, based on data from U.S. Office of Management and Budget and Centers for Disease Control & Prevention, CC BY 4.0. <http://www.tylervigen.com/spurious-correlations>

Chapter 17

1. Newell B (2012) Science alone won't change climate opinions, but it matters. *The Conversation*, 28 November.
2. Spring J (1999) Jackson Pollock, Superstar. *New England Review*. Middlebury College Publications. 20(1).
3. Browning S (2016) All bad poetry is sincere. *The Odyssey Online,* 18 April. <https://www.theodysseyonline.com/all-bad-poetry-is-sincere>
4. Kreuger JI (2016) How not to believe. *Psychology Today*, 13 November. <https://www.psychologytoday.com/us/blog/one-among-many/201611/how-not-believe>
5. Sathyanarayana Rao TS, Andrade C (2011) The MMR vaccine and autism: sensation, refutation, retraction, and fraud. *Indian Journal of Psychiatry* 53(2), 95–96.
6. Dastagir AE (2017) People trust science. So why don't they believe it? *USA Today*, 20 April.
7. National Science Board (2002) *Science and Engineering Indicators 2002*. Division of Science Resource Statistics, National Science Foundation.
8. Field CD (2010) A pot-pourri of beliefs. *British Religion in Numbers*, 2 October. <http://www.brin.ac.uk/a-pot-pourri-of-beliefs/>
9. MacLennan A, Myers S, Taylor A (2006) The continuing use of complementary and alternative medicine in South Australia: costs and beliefs in 2004, *Medical Journal of Australia* 184(1), 27–31.
10. Isaacson W (2011) *Steve Jobs*. Vintage Books, USA.
11. Ronald P (2011) Why parents fear the needle (and the gene). Tomorrow's Table, Science Blogs, 22 January. <https://scienceblogs.com/tomorrowstable/2011/01/22/op-ed-contributor-why-parents>
12. Hull B, Lawrence G, MacIntrye CR, McIntyre P (2002) 'Immunisation coverage Australia 2001'. National Centre for Immunisation Research and Surveillance, University of Sydney.
13. Kahneman D (2011) *Thinking, Fast and Slow*. Penguin, USA.
14. Kahan D (2008) Cultural cognition as a conception of the cultural theory of risk. Cultural Cognition Project Working Paper No. 73.
15. Frazer C (2003) Bridging the gap between the science of childhood immunisation and parents, Vols. 1 & 2. PhD thesis, Australian National University.
16. Nyhan B, Reifler J (2010) When corrections fail: the persistence of political misperceptions. *Political Behavior* 32(2), 303–330.
17. Binder A, Scheufele D, Brossard D, Gunther AC (2010) Interpersonal amplification of risk? Citizen discussions and their impact on perceptions of risks and benefits of a biological research facility. *Risk Analysis* 31(2), 324–334.
18. Ropeik D (2011) Risk perception in toxicology – Part I: Moving beyond scientific instincts to understand risk perception. *Toxicological Sciences* 121(1), 1–6.
19. Parsons J (2018) Latest outrageous 'Flat Earth' theory says Australia is FAKE – and if you think you've been there, you're wrong. *The Mirror*, 3 May.
20. Hafford M (2017) This conspiracy theory proves that Australia isn't real. Refinery29, 24 March. <https://www.refinery29.com/en-us/2017/03/146852/australia-fake-conspiracy-theory>
21. Ronson J (2001) *Them: Adventures with Extremists*. Picador, London.
22. Losh SC (2007) Age, generation and American adult support for pseudoscience. Meetings of the American Association for the Advancement of Science, San Francisco, 18 February.
23. Bridgstock M (2007) Critical thinking: making it popular at university. <www.undeceivingourselves.org>
24. McCrindle Research (2011) <www.mcrindle.com>

25. Sagan C (1996) *The Demon Haunted World: Science as a Candle in the Dark*. Random House, New York.
26. Damisch L, Stoberock B, Mussweiler T (2010) Keep your fingers crossed! How superstition improves performance. *Psychological Science* **21**(7), 1014–1020.
27. McRae M (2011) *Tribal Science: Brains, Beliefs and Bad Ideas*. University of Queensland Press, Brisbane.
28. Jones A (2011) 2GB, 15 March. Cited in Brown MJI (2012) Straw man science: keeping climate simple, *The Conversation*, 27 November.
29. Brown MJI (2012) Straw man science: keeping climate simple. *The Conversation*, 27 November.
30. ABC (2016) Pauline Hanson visits healthy reef to dispute the effects of climate change. ABC News, 25 November.
31. Cook J (2017) What do gorilla suits and blow fish fallacies have to do with climate change? *The Conversation*, 10 February.

Chapter 18

1. Hooker C, Capon A, Leask J (2017) Communicating about risk: strategies for situations where public concern is high but the risk is low. *Public Health Research Practice* **27**(1), e2711709.
2. Covello V, Sandman P (2001) Risk communication: evolution and revolution. In *Solutions to an Environment in Peril*. (Ed. A Wolbarst) John Hopkins University Press, Baltimore, Maryland.
3. Bennett P, Calman K (Eds) (2010) *Risk Communication and Public Health*. Oxford University Press, Oxford and New York.
4. Sandman P (2001) Explaining environmental risk. *The Peter M. Sandman Risk Communication Website*. <http://www.psandman.com/articles/explain3.htm>
5. Sandman P (2012) 'Responding to community outrage: strategies for effective risk communication.' American Industrial Hygiene Association, USA.
6. Sandman P (2002) Four kinds of risk communication, *The Synergist*, April.
7. Benson B (2016), Cognitive bias cheat sheet <https://betterhumans.coach.me/cognitive-bias-cheat-sheet-55a472476b18>
8. Covello VT, Milligan PA (2010) 'Risk communication: principles, tools, & techniques.' United States Nuclear Regulatory Commission, Rockville, Maryland.
9. Fischhoff B, Brewer NT, Downs JS (2011) *Communicating Risks and Benefits: An Evidence-Based User's Guide*. FDA, Department of Health and Human Services, Silver Spring.
10. Sandman P (2010) Empathetic risk communications in high stress situations. *The Peter M. Sandman Risk Communication Website*. <http://www.psandman.com/col/empathy2.htm>
11. Kortenkamp KV, Basten B (2015) Environmental science in the media: effects of opposing viewpoints on risk and uncertainty perceptions. *Science Communication* **37**(3), 287–313.
12. Centers for Disease Control and Prevention (2014) 'Working with the media'. Crisis and Emergency Risk Communication, Centers for Disease Control and Prevention, US Department of Health and Human Services.
13. Cormick C (2011) Risk communication, *Issues Magazine*, December.
14. Gluckman P (2016) Making decisions in the face of uncertainty: understanding risk. Office of the Prime Minister's Chief Science Advisor, New Zealand.

Chapter 19

1. Thaler A (undated) When I talk about climate change, I don't talk about science. *SouthernFriedScience*, <http://www.southernfriedscience.com/when-i-talk-about-climate-change-i-dont-talk-about-science/>
2. National Academies of Sciences, Engineering, and Medicine (2017) *Communicating Science Effectively: A Research Agenda*. The National Academies Press, Washington DC. p. 56.
3. Douglas M, Wildavsky A (1982) *Risk and Culture*, University of California Press, California.
4. Kahan D (2012) Cultural cognition as a conception of the cultural theory of risk. In *Handbook of Risk Theory: Epistemology, Decision Theory, Ethics and Social Implications of Risk* (Eds R Hillerbrand, P Sandin, S Roeser and M Peterson) Springer.

5. Kahan D (2018) Cultural cognition dictionary. *Cultural Cognition Blog* <http://www.culturalcognition.net/cultural-cognition-dictionaryg/>

Chapter 20

1. EUFIC (2014) Motivating behaviour change European Food Information Council Review, 1 July. <https://www.eufic.org/en/healthy-living/article/motivating-behaviour-change>
2. Ferrier A, Fleming J (2014) *The Advertising Effect: How to Change Behaviour.* Oxford University Press, Oxford.
3. Solof M (2010) Travelers behaving badly: behavioral economics offers insights and strategies for improving transportation. *Intransition magazine*, Spring/Summer.
4. Moore J (2018) 10 amazing examples of nudge theory. Clickworks, 29 January. <https://clickworks.ie/10-examples-nudge-theory/>
5. Halpern D (2015) *Inside the Nudge Unit: How Small Changes Can Make a Big Difference.* Random House.
6. Cobern MK, Porter BE, Leeming FC, Dwyer WO (1995) The effect of commitment on adoption and diffusion of grass cycling. *Environment and Behaviour* **27**(2), 213–232.
7. Heimlich J, Ardoin N (2008) Understanding behaviour to understand behaviour change: a literature review. *Environmental Education Research* **14**(3), 215–237.
8. Monroe M (2003) Two avenues for encouraging conservation behaviours. *Human Ecology Review* **10**(2), 113–125.
9. Wittig AF, Belkin G (1990) *Introduction to Psychology.* McGraw-Hill, New York.
10. Willis R, Stewart RA, Panuwatwanich K, Williams PR, Hollingsworth AL (2011) Quantifying the influence of environmental and water conservation attitudes on household end use water consumption. *Journal of Environmental Management* **92**(8), 1996–2009.
11. Dolnicar S, Hurlimann A (2009) Drinking water from alternative water sources: differences in beliefs, social norms and factors of perceived behavioural control across eight Australian locations, *Water Science and Technology* **60**(6), 1433–1444.
12. Bell PA, Greene TC, Fisher J, Baum A (1996) *Environmental Psychology.* Harcourt Brace, Fort Worth, Texas.
13. Ariely D (2009) Irrationality is the real invisible hand. *Psychology Today*, 8 May. <https://www.psychologytoday.com/au/blog/predictably-irrational/200905/irrationality-is-the-real-invisible-hand>
14. Thaler RH (2008) *Nudge.* Yale University Press, New Haven, Conneticut.
15. Kahneman D (2011) *Thinking, Fast and Slow,* Penguin, USA.
16. Office of the Press Secretary (2015) Executive Order – Using behavioral science insights to better serve the American people. The White House, 15 September.
17. Frederiks ER, Stenner K, Hobman EV (2015) Household energy use: applying behavioural economics to understand consumer decision-making and behaviour. *Renewable and Sustainable Energy Reviews* **41**, 1385–1394.
18. Corner A, Randall A (2011) Selling climate change? The limitations of social marketing as a strategy in climate change public engagement. *Global Environment Change* **21**(3), 1005–1014.
19. Jain P (2013) 5 Behavioural economics that marketers can't afford to ignore, *Forbes Tech*, 1 March.
20. Spiegler MD, Guevremont DC (2003) *Contemporary Behaviour Therapy,* 4th edn. Wadsworth Publishing Company, Belomont, California.
21. Eriksen C (2010) Playing with fire? Bushfire and everyday life in changing rural landscapes. PhD thesis, School of Earth and Environmental Sciences, Faculty of Science, University of Wollongong.

Chapter 21

1. Davies L (2012) Jailing of Italian seismologists leaves scientific community in shock, *The Guardian*, 24 October. <https://www.theguardian.com/world/2012/oct/23/jailing-italian-seismologists-scientific-community>

2. National Academies of Sciences, Engineering, and Medicine (2017) *Communicating Science Effectively: A Research Agenda*. The National Academies Press, Washington DC.

3. Nelkin D (Ed.) (1991) *Controversy: Politics of Technical Decisions*, SAGE.

4. Easterbrook C, Maddern G (2008) Porcine and bovine surgical products: Jewish, Muslim, and Hindu perspectives. *Archives of Surgery* **143**(4), 366–370.

5. Pew Research Center (2015) 'Religion and science'. Pew Research Center, Washington DC.

6. Harker D (2015) *Creating Scientific Controversies*. Cambridge University Press, UK.

7. Armitage C (2013) Sunscreen fear 'a risk to health', *Sydney Morning Herald*, 1 April.

8. Prothero D (2014) The holocaust denier's playbook and the tobacco smokescreen common threads in the thinking and tactics of denialists and pseudoscientists. In *Philosophy of Pseudoscience: Reconsidering the Demarcation Problem* (Eds M Pigliucci and M Boudry), Chicago Scholarship Online.

9. Black C, Fielding-Smith A (2017) Astroturfing, Twitter bots, amplification: Inside the online influence industry, *New Statesman Tech*, 7 December.

10. Tapper J, Culhane M (2006) Al Gore YouTube spoof not so amateurish. ABC News, 4 August. <https://abcnews.go.com/GMA/story?id=2273111&page=1>

11. Oreskes N, Conway E (2010) *Merchants of Doubt: How a Handful of Scientists Obscured the Truth on Issues from Tobacco Smoke to Global Warming*. Bloomsbury Press.

12. Madgrigal A (2009) 7 (Crazy) civilian uses for nuclear bombs. *Wired*, 10 April.

13. Boutron I, Altman DG, Hopewell S, Vera-Badillo F, Tannock I, Ravaud P (2014) Impact of spin in the abstracts of articles reporting results of randomized controlled trials in the field of cancer: The SPIIN randomized controlled trial. *Journal of Clinical Oncology* **32**(36), 4120–4126.

14. Blake M (2018) Why bullshit hurts democracy more than lies. *The Conversation*, 14 May.

15. Ding D, Maibach EW, Zhao X, Roser-Renouf C, Leiserowitz A (2011) Support for climate policy and societal action are linked to perceptions about scientific agreement. *Nature Climate Change* **1**(9), 462–466.

16. Engelhardt HT Jr, Caplan A (Eds) (1987) *Scientific Controversies: Case Studies in the Resolution and Closure of Disputes in Science and Technology*. Cambridge University Press, Cambridge UK.

Chapter 22

1. Handwerk B (2017) Friday the 13th is back. Here's why it scares us. *National Geographic*, 10 October.

2. Statista (2018) Superstition: Do you believe the following, or not? statista.com.

3. Garrett BM, Cutting RL (2017) Magical beliefs and discriminating science from pseudoscience in undergraduate professional students. *Heliyon*, 11 November.

4. <https://www.casino.org/superstitious-states/>

5. Damisch L, Stoberock B, Mussweiler T (2010) Keep your fingers crossed! How superstition improves performance. *Psychological Science* **21**(7), 1014–1020.

6. Griffiths BM (2018) pers.comm., 8 May.

7. Hansson SO (2017) Science denial as a form of pseudoscience. *Studies in History and Philosophy of Science Part A* **63**, 39–47.

8. Schultz EN (2015) Pseudoscience and conspiracy theory are not victimless crimes against science. *The Conversation*, 4 June.

9. Campbell H (2014) The food babe took down her goofy microwave oven post – science win. *Science 2.0.* 20 July.

10. dÉntremont Y (2015) The 'Food Babe' blogger is full of shit, 6 April. Gawker.com.

11. Gorksi D (2015) The Wellness Warrior, Jess Ainscough, has passed away. *Respectful Insolence*, 27 February.

12. Townson S (2016) Why people fall for pseudoscience (and how academics can fight back). *The Guardian*, 26 January.

13. Cook J, Lewandowsky S (2012) *The Debunking Handbook*. University of Queensland, St Lucia, Australia.
14. Lewandowsky S, Ecker UK, Seifert CM, Schwarz N, Cook J (2012) Misinformation and its correction continued influence and successful debiasing. *Psychological Science in the Public Interest* **13**(3), 106–131.
15. Lewandowsky S (2017) Claiming that Listerine alleviates cold symptoms is false: To repeat or not to repeat the myth during debunking? *Skeptical Science*, 22 June.
16. Bolsen T, Druckman JN (2015) Counteracting the politicization of science. *Journal of Communication* **65**(5), 745–769.

Chapter 23
1. Oreskes N, Conway E (2010) *Merchants of Doubt: How a Handful of Scientists Obscured the Truth on Issues from Tobacco Smoke to Global Warming*. Bloomsbury Press.
2. Keohane RO, Lane M, Oppenheimer M (2014) The ethics of scientific communication under uncertainty. *Politics, Philosophy & Economics* **13**(4), 343–368.
3. National Academies of Sciences, Engineering, and Medicine (2017) *Communicating Science Effectively: A Research Agenda*. The National Academies Press, Washington DC.
4. Nisbet MC, Scheufele DA (2009) What's next for science communication? Promising directions and lingering distractions. *American Journal of Botany* **96**(10), 1767–1778.
5. Bronson K (2018) Excluding 'Anti-biotech' activists from Canadian agri-food policy making: ethical implications of the deficit model of science communication. In *Ethics and Practice in Science Communication* (Eds S Priest, J Goodwin and MF Dahlstrom) University of Chicago Press.
6. Priest S, Goodwin J, Dahlstrom MF (Eds) (2018) *Ethics and Practice in Science Communication*, University of Chicago Press.
7. Nisbet MC (2009) The ethics of framing science. In *Communicating Biological Sciences: Ethical and Metaphorical Dimensions* (Eds B Nerlich, B Larson and R Elliott) Ashgate, London, England.
8. Mak IW, Evaniew N, Ghert M (2014) Lost in translation: animal models and clinical trials in cancer treatment. *American Journal of Translational Research* **6**(2), 114–118.
9. Ioannidis JPA (2005) Why most published research findings are false. *PLoS Medicine* 2(8), e124.
10. Baker M (2016) 1500 scientists lift the lid on reproducibility. *Nature* **533**, 452–454.
11. Lariviere V, Gingras Y, Archambault E (2009) The decline in the concentration of citations, 1900–2007. *Journal of the Association for Information Science and Technology* **60**(4), 858–862.
12. Sumner P, Vivian-Griffiths S, Boivin J, Williams A, Bott L, Adams R, Venetis CA, Whelan L, Hughes, B Chambers CD (2016) Exaggerations and caveats in press releases and health-related science News. *PLoS ONE* **11**(12), e016217.
13. Sumner P, Vivian-Griffiths S, Boivin J, Williams A, Venetis CA, Davies A, Ogden J, Whelan L, Hughes B, Daltonm B, Boy F, Chambers CD (2014) The association between exaggeration in health related science news and academic press releases: retrospective observational study. *British Medical Journal* **349**, g7015.
14. Horton R (2015) Offline: What is medicine's 5 sigma? *The Lancet* **385**(9976), 1380.
15. Flam F (2013) How science {writing} goes wrong: an overreaching critique of the whole of science, *Undark*, 23 October.
16. Gasparyan AY, Yessirkepov M, Diyanova SN, Kitas GD (2015) Publishing ethics and predatory practices: a dilemma for all stakeholders of science communication. *Journal of Korean Medical Science* **30**(8), 1010–1016.
17. Weisberger M (2017) #DoesItFart: database answers your burning questions about animal gas. *LiveScience*, 20 January. <https://www.livescience.com/57565-animal-fart-database.html>
18. Battley P (2018) Kill or cure? <http://kill-or-cure.herokuapp.com/>
19. Fox F, Lethbridge F (2018) A new labelling system for medical research press releases. Science Media Centre, 7 June. <http://www.sciencemediacentre. org/a-new-labelling-system-for-medical-research-press-releases/>

20. Emery D (2017) Did a study show that staring at breasts is good for men's health? *ThoughtCo*, 16 May. <https://www.thoughtco.com/staring-at-breasts-is-good-for-mens-health-3299606>
21. <www.snopes.com>
22. Vosoughi S, Roy D, Aral S (2018) The spread of true and false news online. *Science* **359**(6380), 1146–1151.
23. Bornmann L, Mutz R, Daniel HD (2007) Gender differences in grant peer review: a meta-analysis. *Journal of Infometrics* **1**(3), 226–238.
24. Knobloch-Westerwick S, Glynn CJ (2013) The Matilda effect – role congruity effects on scholarly communication: A citation analysis of communication research and *Journal of Communication* articles. *Communication Research* **40**(1), 3–26.
25. Lincoln AE, Pincus S, Koster JB, Leboy PS (2012) The Matilda effect in science: Awards and prizes in the US, 1990s and 2000s. *Social Studies of Science* **42**(2), 307–320.
26. Trix F, Psenka C (2003) Exploring the color of glass: Letters of recommendation for female and male medical faculty. *Discourse & Society* **14**(2), 191–220.
27. Ledin A, Bornmann L, Gannon F, Wallon GA (2007) A persistent problem: Traditional gender roles hold back female scientists. *EMBO Reports* **8**(11), 982–987.
28. Ramšak A (2014) 'Guidelines for gender sensitive reporting.' Ministry of Foreign Affairs, Republic of Slovenia.
29. Grizzle A (Ed.) (2012) *Gender-Sensitive Indicators for Media Communication and Information Sector.* UNESCO, Paris.
30. Cegnar T, Benestad RE, Billard C (2010) Is there a need for a code of ethics in science communication and communicating uncertainties on climate change? In 10th EMS Annual Meeting, 10th European Conference on Applications of Meteorology (ECAM). 13–17 September, Zurich, Switzerland.
31. Medvecky F, Leach J (2017) The ethics of science communication. *Journal of Science Communication* **16**(04).
32. Medvecky F (2018) Should we talk about the monster? Reconsidering the ethics of science Communication. Lecture, Australian National University, 5 June.
33. Beauchamp T, Childress J (2013) *Principles of Biomedical Ethics.* Oxford University Press, New York.
34. Stauber JC, Rampton S. (1995) *Toxic Sludge Is Good for You! Lies, Damn Lies and the Public Relations Industry.* Common Courage Press, Monroe, Maine.

Chapter 24
1. Kruglinksi S (2006) Whatever happened to cold fusion? *Discover Magazine*, 3 March.
2. Open Science Collaboration (2015) Estimating the reproducibility of psychological science. *Science* **349**, aac4716.
3. Shih M, Pittinsky TL, Ambady N (1999) Stereotype susceptibility: Identity salience and shifts in performance. *Psychological Science* **10**(1), 80–83.
4. Gibson CE, Lose, J, Vitiello C (2014) A replication attempt of stereotype susceptibility: Identity salience and shifts in quantitative performance. *Social Psychology* **45**(3), 194–198.
5. Moon A, Roeder SS (2014) A secondary replication attempt of stereotype susceptibility. *Social Psychology,* **45**(3), 199–201.
6. Vedentam S, Penman M (2016) When great minds think unalike: Inside science's 'Replication Crisis'. *Hidden Brain*, 24 May.
7. Van Bavel JJ, Mende-Siedlecki P, Brady WJ, Reinero D A (2016) Reply to Inbar: Contextual sensitivity helps explain the reproducibility gap between social and cognitive psychology. *Proceedings of the National Academy of Sciences of the United States of America.* **113**(34), E4935–E4936.
8. Van Bavel JJ (2016) Why do so many studies fail to replicate? *New York Times*, 27 May.
9. Watters E (2013) We aren't the world. *Pacific Standard Magazine*, 25 February.
10. Henrich J, Heine SJ, Norenzayan A (2010) The weirdest people in the world? *Behavioral and Brain Sciences* **33**(2–3): 61–83.

11. Stafford T (2018) The backfire effect is elusive. *Mind Hacks*, 3 January. <https://mindhacks.com/2018/01/03/the-backfire-effect-is-elusive/>

12. Nyhan B, Reifler J (2010) When corrections fail: The persistence of political misperceptions. *Political Behavior*, **32**(2), 303–330.

13. Engber D (2018) LOL something matters. *Slate*, 3 January. <https://slate.com/health-and-science/2018/01/weve-been-told-were-living-in-a-post-truth-age-dont-believe-it.html

14. Van Heuvelen B (2007) The Internet is making us stupider. *Salon*, 7 November. <https://www.salon.com/2007/11/07/sunstein/>

15. Wood T, Porter E (2019) The elusive backfire effect: Mass attitudes' steadfast factual adherence. *Political Behavior* **41**, 135–163.

16. Guess AM (2016) Media choice and moderation: Evidence from online tracking data. New York University, 7 October. (unpublished)

17. Nelson JL, Webster JG (2017) The myth of partisan selective exposure: A portrait of the online political news audience. *Social Media and Society* **3**(3), 1–13.

18. Oremus W (2017) The filter bubble revisited. *Slate*, 5 April. <https://slate.com/technology/2017/04/filter-bubbles-revisited-the-internet-may-not-be-driving-political-polarization.html>

Chapter 25

1. Riedlinger M (2017) The world of the science communication practitioner (conference presentation), Rockefeller Science Communication Conference, 20 December.

Index